THE
NEW
Microbiology
From Microbiomes to CRISPR

THE
NEW
Microbiology

From Microbiomes to CRISPR

Pascale Cossart
Institut Pasteur
Paris, France

ASM
PRESS
WASHINGTON, DC

Cover image: *Bacillus subtilis* bacteria in the process of sporulation. The spore membrane is green and the bacterial membrane is red. Courtesy of Javier Lopez Garrido and Kit Pogliano, University of California, San Diego.

Translation of *La nouvelle microbiologie: Des microbiotes aux CRISPR* by Pascale Cossart, (C) ODILE JACOB, 2016

Library of Congress Cataloging-in-Publication Data

Names: Cossart, Pascale, author.
Title: The new microbiology : from microbiomes to CRISPR / Pascale Cossart, Institut Pasteur, Paris, France.
Other titles: Nouvelle microbiologie. English
Description: Washington, DC : ASM Press, [2018] | Includes bibliographical references and index.
Identifiers: LCCN 2018001322 (print) | LCCN 2018003826 (ebook) | ISBN 9781683670117 (e-book) | ISBN 9781683670100 | ISBN 9781683670100 (print)
Subjects: LCSH: Microbiology.
Classification: LCC QR41.2 (ebook) | LCC QR41.2 .C6713 2018 (print) | DDC 579—dc23
LC record available at https://lccn.loc.gov/2018001322

doi:10.1128/9781683670117

Printed in the United States of America

10 9 8 7 6 5 4 3 2 1

Address editorial correspondence to: ASM Press, 1752 N St., NW, Washington, DC 20036-2904, USA.
Send orders to: ASM Press, P.O. Box 605, Herndon, VA 20172, USA.
Phone: 800-546-2416; 703-661-1593. Fax: 703-661-1501.
E-mail: books@asmusa.org
Online: http://www.asmscience.org

CONTENTS

PART III:
The Biology of Infections

PART IV:
Bacteria as Tools

PREFACE

For many of us, the word "microbe" still conjures a negative image, one of sickness, infection, or contamination. In general, we do not wonder "Where does this microbe come from?" except in case of an epidemic. We simply observe that its presence is inopportune and is dismantling a preestablished order, an equilibrium: this well-being that is named "health."

We now know, however, that good health depends on the presence of millions of beneficial microbes and microorganisms. These live on our skin and in different places in our bodies, such as the intestine, the mouth, and the nose, or they participate in various processes, such as the making of cheese, yogurt, and other foods, or water treatment and environment decontamination. They play a key role in maintaining the stability of our environment and the biodiversity of the flora and fauna of our planet.

Thanks to the studies of Louis Pasteur and Robert Koch at the end of the 19th century, it is well established that microbes do not spontaneously generate, that each microbe is born from another microbe, and that the smallest living organisms capable of autonomous life are called bacteria (from the Greek *baktēria*, meaning a stick or rod, named for the rod-like shape of the first observed bacteria). These bacteria, observable with simple microscopes, are single-celled organisms that can generate thousands of similar unicellular daughter cells.

Louis Pasteur and Robert Koch importantly discovered that bacteria were responsible for numerous diseases that have devastated humanity for thousands of years, such as the plague, cholera, and tuberculosis. Their studies paved the way for powerful methods of diagnosis of, and treatment for, bacterial infections, and for the development of vaccines, some of which are still being used today. Pasteur and Koch also introduced the concept of the study of bacteria in general, whatever their nature—i.e., either pathogenic, illness-generating bacteria or nonpathogenic bacteria that carry out other functions. In fact, the discoveries of Pasteur, Koch, and their collaborators were so revolutionary and so important that by the early 20th century they triggered an immense interest, first among medical doctors and then among biologists of all sorts attracted to this new discipline: microbiology, the study of various microorganisms invisible to the naked eye, and more specifically, bacteriology, the study of bacteria.

During this flourishing period and the entire century that has followed, the field has advanced by leaps and bounds in many directions. At first, shortly after Pasteur and Koch, microbiology developed rather slowly, with the meticulous identification of all kinds of bacteria, the establishment of various collections, and diverse classifications and precise descriptions. Then things really sped up. In the early 1950s, the discovery of DNA as the basis of the genetic material of all living organisms, combined with the previous research on bacteria, quickly led to the development of concepts that applied, as Nobel laureate Jacques Monod put it, to the bacterium as much as to the elephant. These concepts included DNA replication, DNA transcription, protein translation, and protein synthesis. This in turn led to the development of molecular biology and genetic engineering: the art of manipulating genes and species.

By the end of the 20th century, technologies in DNA sequencing—the determination of the structure of genes and, soon, of complete bacterial genomes—sparked a totally unexpected acceleration in the study of bacteria, both pathogenic and not. Our understanding of infectious diseases was completely redefined by these approaches that, in association with cellular biology techniques such as imaging, started to shed light on the multiple mechanisms used by microorganisms to establish infection by interacting in various ways with the infected host and by harnessing many of the host's essential functions and fundamental mechanisms.

In parallel with this new vision on infectious diseases, research on the behavior of bacteria has shown that all bacteria without exception have a social life. They can live in small groups and diverse communities known as *biofilms* present on all kinds of surfaces. They can live in harmony with their fellow bacteria in heterogeneous, but stable, groups. When these groups grow in size and associate with other microorganisms, including parasites or viruses, they are called *microbiomes*. What was once known as the "intestinal flora" is now termed the *intestinal microbiome*. The intestinal microbiome is not the only type of microbiome; other parts of the body, and other organisms, feature their own. We now know that these microbiomes evolve and that they are unique to the individual they inhabit, based on their host's specific eating habits, genetic heritage, underlying illnesses, and even personal behavior.

Even if bacteria seem to live independently in nature, many exist in symbiotic relationships not only with humans but also with all animals, including insects, and even plants. This cohabitation sometimes produces stunning effects on the host, such as sterility and even the eradication of males in insects. Bacteria present on plant roots can help them capture the soil nitrogen essential for the plant's growth.

Bacteria have very elaborate social lives. In addition to their ability to live in groups, and in order to do so, they can communicate using a chemical language that allows them to recognize and distinguish one another by species or family. Bacteria use these chemical languages to cooperate against a common enemy. For example, some pathogenic bacteria will not deploy their attack mechanisms unless they are numerous enough to succeed. Some bacteria can also regulate the times when they become luminescent, lighting up only once their numbers reach a certain threshold.

In order to adapt to various situations and to decide when to use their special capabilities, bacteria employ very sophisticated regulatory mechanisms. Each bacterial component, from proteins to small molecules, including vitamins and metals, participates in multiple adaptation mechanisms that bacteria put into action at various points in their lives. The molecules that participate in the controlled expression of genomes, and on which researchers have made the most progress recently, are RNA molecules. François Jacob and Jacques Monod hypothesized that RNAs could regulate gene expression, but they never imagined that RNAs could regulate gene expression in so many different ways. Bacterial RNA, considered as recently as the end of the last century to be mostly a production intermediary between DNA and proteins (hence the term *messenger RNA*), plays various and sometimes surprising roles. One of the most important recent advances in biology is the discovery that bacteria have extremely effective RNA-dependent defense strategies in place, known as CRISPR (pronounced *crisper*) for clustered regularly interspaced palindromic repeats, which they use to protect themselves from the bacteria-infecting viruses known as *bacteriophages*, or just *phages*. Specifically, bacteria remember their first encounter with a given phage and are able to put in place a kind of immunity, "vaccinating" themselves against this phage.

These bacterial systems work so well and are so adaptable that they are now the basis for a revolutionary technique, the CRISPR/Cas9 technology, that allows genome editing in all organisms that have been tested so far. This method makes genome modification quick and easy, and the mutations created allow for sophisticated studies of gene function or for the replacement of defective genes, paving the way for gene therapies. The CRISPR/Cas9 technology was recognized by a Breakthrough Prize in Life Sciences in 2015 in the United States and by numerous other prestigious international prizes that honor great scientific advances.

Bacteria defend themselves not only from viruses but also from their fellow bacteria, which are sometimes very aggressive. To do this, they produce many kinds of toxins and antibacterial poisons for which they themselves have one or more immunity proteins. In the bacterial world, the struggle for life is continually taking place on an infinitesimal scale. But could these antibacterial poisons also be used on a much larger scale, to fight and gain better control over pathogenic bacteria? They certainly constitute a foreseeable strategy for replacing antibiotics that have become ineffective.

In fact, antibiotics have been, for decades, the most used antibacterial agents. Unfortunately, bacteria have adapted accordingly, developing resistances that have dramatic medical consequences, as in the case of the bacterium responsible for tuberculosis (*Mycobacterium tuberculosis*, or the Koch bacillus). We are no longer able to treat certain serious illnesses, and, as a result, they are coming back with a vengeance. The alarm has been sounded. The public is aware that this is a worldwide concern. Nevertheless, there are now reasons for optimism, or at least hope. Based on our recent knowledge, we are discovering new, alternative ways of fighting pathogens, raising new hopes for more effective treatments. For example, we can use our knowledge of bacterial genomes to identify inhibitors of chemical reactions or metabolic pathways that exist only in bacteria, not in humans.

Nevertheless, the threat of returning to a "preantibiotic" era is real and must be taken into account. We must therefore maintain constant vigilance when putting in place new therapies or when halting formerly obligatory vaccinations. Would it be reasonable, for example, to continue the policy in France of restricting vaccination with BCG (bacille Calmette-Guérin) against tuberculosis? Such questions should be carefully considered, especially in our global society where travel to and from countries with lower vaccination rates can be easy.

The objective in this book is to illustrate that very important discoveries and new concepts have come to light in the last few decades. These developments clearly show that the field of microbiology has undergone a *bona fide* revolution and that the amazing renaissance that is taking place can have wide-ranging consequences. This new understanding is going to change our daily lives dramatically, from our eating habits and daily routines to our way of looking at the rest of the living organisms on Earth: bacteria, plants, animals, even insects. In addition, recent discoveries will help us implement new strategies for fighting pathogenic agents and battle not only infectious diseases, but also their vectors. An example already in place in Australia is a plan to eliminate certain disease-bearing mosquitoes by releasing into the wild male mosquitoes that have been rendered sterile by infection with *Wolbachia* bacteria.

This book is limited to the rebirth of bacteriology, in part because this is my own domain of expertise. That said, virology, parasitology, and mycology are also mentioned because these areas benefit from the same technological advances. Bacteriology, however, is the field that has been the most profoundly impacted by these advances and that consequently has benefited from the development of the greatest number of new concepts.

It was predicted that the 21st century would be the age of biology. This is indeed the case, and microbiology is at the forefront. In 2012, the French Academy of Sciences, with its sister institutions in England and Germany, the Royal Society and the Leopoldina, held a colloquium titled "The New Microbiology" that met with great success. I have used the same title for this book.

ACKNOWLEDGMENTS

I would like to sincerely thank my fellow collaborators Olivier Dussurget and Nathalie Rolhion for their meticulous critical reading of this manuscript, as well as Carla Saleh and Didier Mazel. Thanks to Juan J. Quereda for finalizing the drawings and the figures; Urs Jenal and Javier Lopez Garrido for the images of *Caulobacter* and *Bacillus subtilis* (cover photo); Caroline Dean, Jean-Pierre Caillaudeau, and Bruno Lemaître for their suggestions; and Nicolas Witkowski for all his patience and advice.

I am grateful to Chloe A. M. Hagen for the primary translation of the French text, and to Megan Angelini and Ellie Tupper for their skilled and careful editing to bring the spirit of the French edition to English speakers. I sincerely thank Christine Charlip for her careful and meticulous coordination of the English edition as well as Greg Payne for his enthusiasm for publishing the book.

Finally, thanks to Odile Jacob for her enthusiasm and the pleasure of our discussions on this new microbiology!

ABOUT THE AUTHOR

After studying chemistry in Lille, France, Pascale Cossart obtained a masters degree from Georgetown University, Washington, DC. Returning to France, she obtained her Ph.D. in Paris at the Institut Pasteur, where she now heads the Bacteria-Cell Interactions unit, which is also an Inserm and an INRA unit. After studying DNA-protein interactions, in 1986 Dr. Cossart began to study the molecular and cellular basis of infections by intracellular bacteria, taking as a model the bacterium *Listeria monocytogenes.* Her research has led to new concepts in infection biology as well as in microbiology, cell biology, and epigenetics.

Pascale Cossart is considered a pioneer in cellular microbiology. Her contributions have been recognized by a number of international awards, including the L'Oreal/UNESCO Prize for Women in Science (1998), the Richard Lounsbery Prize (Académie des Sciences, Paris/National Academy of Sciences, 1998), the Robert Koch Prize (2007), the Louis Jeantet Prize for Medicine (2008), and the Balzan Prize (2013). She is a member of the French Academy of Science (2002); a foreign member of the National Academy of Sciences (United States; 2009), the German Leopoldina (2001), the Royal Society (United Kingdom; 2010), and the National Academy of Medicine (United States; 2014); and, since January 2016, Secrétaire Perpétuel de l'Académie des Sciences, Institut Pasteur. (Photo courtesy of Agnès Ullmann.)

PART I

New Concepts in Microbiology

CHAPTER 1

Bacteria: Many Friends, Few Enemies

Bacteria are unicellular living organisms that make up one of the three domains of life: *Bacteria*, *Archaea*, and *Eukaryota* (Fig. 1). This model of three branches stemming from a common ancestor was first proposed by Carl Woese in 1977. The absence of a nucleus is one major difference between prokaryotes and eukaryotes. Eukaryota or eukaryotes include animals, plants, fungi, and protozoa, which all have nuclei; bacteria and archaea are prokaryotes and do not have a nucleus. The DNA of prokaryotes is non-membrane bound, unlike in eukaryotes. But do not assume that bacteria are merely small sacks full of disorderly contents. Their "interior" is in fact very well organized.

Archaea, like bacteria, are unicellular organisms but differ from bacteria in that they have lipids that are not found in bacteria and an ensemble of compounds that are similar to those of eukaryotes, in particular the machinery that regulates gene expression. When they were discovered, archaea were thought to exist only in extreme environments, such as very

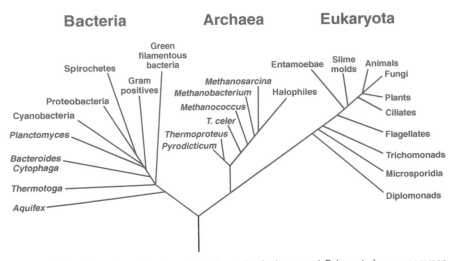

Figure 1. The three large domains of life. *Bacteria*, *Archaea*, and *Eukaryota* have a common ancestor.

hot water springs, but we now know that they are present everywhere, including in our gut.

Bacteria are extremely varied and make up the most diverse domain of life. They have been on Earth for billions of years and have evolved to survive in a great variety of conditions. There are more than 11,500 known species of bacteria in more than 2,000 genera (groupings of species). These numbers have so far been based only on gene comparisons, particularly the 16S RNA genes, and they keep rising. Classification methods are changing too. Now that we can compare entire genome sequences, the definition of "species" itself is evolving.

Bacteria may have different shapes (Fig. 2). There are four main categories: cocci, or spheres; bacilli, or rods; spirals; and comma-shaped, or curved bacteria. All bacteria divide, regardless of their shape. One bacterium splits into two, via an asexual reproduction. Nevertheless, genetic material can be exchanged between two bacteria by means of mechanisms described as *horizontal gene transfer*. We will come back to this topic later on.

Bacteria are present everywhere in the environment; they are in all habitats on earth, including hot springs and seawater, even in very high salinity. Many live in humans—there are an estimated 10^{10} bacteria on our skin, 10^{10} in our mouths, and 10^{14} in our intestines. That's 10 times more bacterial cells in our bodies than human cells! However, a recently published article investigated this number and concluded that it was actually overestimated by a factor of 10. Whatever the count, in our intestine—which contains tens

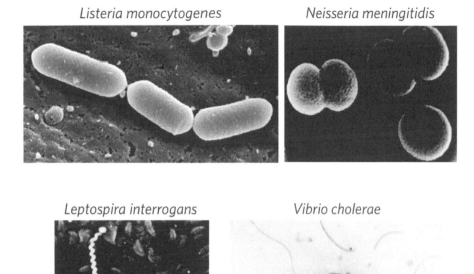

Listeria monocytogenes   Neisseria meningitidis

Leptospira interrogans   Vibrio cholerae

Figure 2. The four main types of bacteria: bacilli (*Listeria monocytogenes*), cocci (*Neisseria meningitidis* or meningococci), spirals (*Leptospira interrogans*), and comma-shaped (*Vibrio cholerae*).

of billions of bacteria—there are more than a thousand different species. Sometimes I like to think that bacteria are our constant companions, generally friendly hitchhikers that we carry around with us everywhere!

Bacteria first appeared more than 3 billion years ago—that's 2 billion years before animals—and have since lived more or less undisturbed in the biosphere. We do not know for sure how the first organism with a nucleus was born, but it was probably from the fusion of a bacterium with an archaeon. Indeed, genes from both of these domains are present in animals; it is clear that an ancestor of all modern eukaryotes must have "swallowed" a bacterium, leading to the stable symbiotic relationship that produced the energy-producing compartments called mitochondria in all of our cells. These small organelles somewhat resemble bacteria and are indispensable to the formation of thousands of compounds, most notably ATP, a chemical compound that temporarily stores energy and is used for many chemical reactions in cells. One could say that the first animals started out as bacterivores before they became herbivores, carnivores, or omnivores!

Many bacteria live free in natural environments. There they live, grow, and feed and by doing so contribute to the equilibrium and characteristics of the specific ecosystems in which they grow. For example, bacteria from the *Streptomyces* family are responsible for the so refreshing smell of the woods after a rain.

Many other bacteria are not alone and are associated with a partner. They establish long-lasting relationships that are mutually beneficial, or "symbiotic," within humans, animals, and even plants. Additionally, as we will soon see, bacteria of several species can assemble in very large communities, called *microbiomes*, which become integral part of organisms. These combinations of organisms and microbiomes are referred to as *superorganisms*.

It is important to realize that of all the bacteria on Earth, pathogenic bacteria (those responsible for disease) are in the minority. Among those, a few produce very powerful toxins and always induce disease. An example is *Vibrio cholerae*, the water-transmissible *Vibrio* species responsible for cholera, which produces a toxin that can cause fatal diarrhea and dehydration. Another example is *Corynebacterium diphtheriae*, which causes diphtheria, a disease nearly forgotten in developed countries due to the efficacy of mandatory vaccinations. Also in this group is *Clostridium tetani*, which produces the toxin for tetanus, and *Clostridium botulinum*, which produces the botulinum toxin.

However, illnesses caused by a single toxin are extremely rare. As a general rule, bacteria create disease by means of an arsenal of strategies and tools called *virulence mechanisms* and *virulence factors*. The combination of different virulence factors allows a bacterium to enter an organism, evade the host's defense mechanisms, and multiply and invade different parts of the body such as the throat (*Streptococcus*), the lungs (*Legionella*), the intestines (*Salmonella*), or the nose and pharynx (*Pneumococcus*). Bacteria often can establish an infection only if the host's immunity is weakened by fatigue by a viral or other infection (such as pneumococcal infections that often follow respiratory tract inflammation due to influenza), by medical treatment (such as immune suppression due to chemotherapy), or by a genetic mutation.

The importance of the genetic context with regard to a host's susceptibility to infection is now the subject of extensive studies by scientists around the world. The laboratory led by Jean-Laurent Casanova is a strong proponent of the hypothesis that disease can be or is linked to the host's genetics. These investigators have found evidence for genetic predisposition to several illnesses, for example, susceptibility to infection with low-pathogenicity mycobacterial strains including BCG, the live attenuated strain of *Mycobacterium bovis* used to vaccinate against tuberculosis. Some children have a defect in one of the genes coding for interferon or for interferon signaling, which can result in unanticipated and sometimes fatal illness after vaccination.

Throughout this book, we will learn that bacteria, both pathogenic and nonpathogenic, are living organisms with a range of diverse properties, some of which are quite unexpected. With their unique capacity to produce vital compounds or to transform or eliminate other compounds, bacteria contribute constantly to the everyday lives of their hosts and to the balance of the ecosystems where they live.

CHAPTER 2

Bacteria: Highly Organized Unicellular Organisms

Bacteria seem to be very simple cells, having neither a nucleus nor internal organelles as in animal or plant cells, but they actually have a highly structured internal organization. Their shape is very precise and their content is well organized. Each protein or group of proteins has its specific location, from one generation to the next.

Most bacteria have an outer layer consisting of a thin membrane that is covered by a cell wall made up of *peptidoglycan*, which gives each bacterium its shape and rigidity and allows it to survive drastic differences between its interior and its exterior environment, including differences in temperature, pH, and salinity. Some bacteria also have a second external membrane and also a capsule.

Bacteria are often classified into two categories, Gram positive and Gram negative. These terms derive from the staining technique developed by Hans Christian Gram, in which a purple dye binds to the peptidoglycan in bacteria that have a thick cell wall but no external membranes and is negative in all other

Mycoplasma

Mycoplasmas, of the genus *Mycoplasma*, are bacteria that have cell membranes but lack a cell wall. These relatively small bacteria have historically been difficult to locate and identify because the principal bacterial identification method, the Gram stain, is based on identification of the peptidoglycan in bacterial cell walls. Our inability to detect *Mycoplasma* has been a problem because many mycoplasmas, which are in most cases *commensals* (i.e., nonpathogenic) in the respiratory tract or vagina, can be responsible for sexually transmitted infections. Also, because these organisms have no cell walls, they are not affected by antibiotics that target peptidoglycan. Mycoplasmas have some of the smallest bacterial genomes, and the chromosome of *Mycoplasma genitalium* was the first to be synthesized by means of synthetic biology tools.

Flagella and other appendages

Flagella are found on the surface of bacteria. They are long helical filaments connected to small rotary motors that allow the bacteria to move and spread out in fluid environments. Hairlike appendages known as *pili* are also found on the surface of some bacteria. Pili allow bacteria to adhere to biological and abiotic surfaces or even to aggregate. *Curli*, similar to pili, are involved in adherence to surfaces and bacterial aggregation. They closely resemble the aggregating "amyloid" fibers found in the brains of Alzheimer patients.

cases. The Gram stain uses crystal violet, a purple stain that binds to the peptidoglycan in bacterial cell walls. The process concludes with a wash with a decolorizing agent and counterstain. Bacteria that have cell walls but no external membranes retain the stain (yielding a purple-colored "positive" cell) while bacteria with external membranes and a thinner cell wall do not retain the crystal violet (yielding pink-colored "negative" cells).

The shape of bacteria is maintained by their cell wall, but their size depends on the volume of their interior, which expands during bacterial growth. Bacteria have a highly organized internal architecture thanks to molecules that are similar to those found in human or plant cells. It was discovered that bacteria even have skeletons, spiral-shaped *cytoskeletons* made of actin that are involved in the localization and activation of enzymes that create peptidoglycan and are crucial for cell division.

Bacterial shape and division: proteins similar to actin and tubulin in eukaryotes

Bacillus-type (rod-shaped) bacteria and their cell walls grow depending on the growth medium composition. Growth of the peptidoglycan, and thus also of the cell wall, is controlled by MreB, a protein anchored to the membrane that is similar to the protein actin in eukaryotes. MreB forms a sort of spiral structure that gives bacilli their elongated form. A protein known as crescentin is responsible for the crescent shape of *Caulobacter* cells. Bacterial growth has a limit. Once a bacterium reaches a certain size, it splits. This is a highly precise process that uses at least two other molecules similar to actin and tubulin, proteins that were once thought to be present only in eukaryotes. In addition, when a bacterium divides, every vital element of the bacterium, including the chromosome (more specifically the DNA), is duplicated and shared between the two resulting daughter cells. At the final phase of division, the key step in the separation of two daughter cells involves another actin-like protein, FtsA, which attaches to the site of division, FtsZ, a protein similar to the tubulin protein in eukaryotic cells. ParM is another protein similar to eukaryotic actin that is involved in the distribution of plasmid DNA between daughter cells.

Most bacteria produce two apparently identical daughter cells. However, some, such as *Caulobacter crescentus*, do not. *Caulobacter* is an aquatic bacterium that has become an impressive model for asymmetrical cell division (Fig. 3). A *Caulobacter* cell about to divide is immobile, attached by a short stem to a surface such as a rock or the ocean bottom. The unattached end generates a mobile daughter cell that uses a flagellum to propel itself away from the still-attached daughter cell, which will continue to grow and split off more daughters. The mobile cell eventually loses its flagellum and grows a stem that it uses to attach itself to a surface, where it will grow and eventually split off daughter cells of its own.

Are bacteria immortal? Do they have strategies for survival? When stressed by conditions such as desiccation or nutrient deficiency, some bacteria reproduce by forming *spores*. These are a type of dormant cell that is extremely resistant to heat, cold, dryness, and even some antiseptics. Spores allow bacteria to persist for years or even centuries—and to disseminate (Fig. 4). When a spore arrives in an appropriate environment, it can germinate and resume a normal binary cell division.

Not all bacteria produce spores, but several that do are considered some of the most dangerous to humans. In late 2001 in the United States, anthrax spores (spores of *Bacillus anthracis*) sent through the mail as an act of bioterrorism caused skin, intestinal, and lung infections that led to the deaths

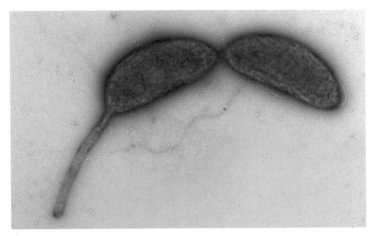

Figure 3. *Caulobacter crescentus* is used as a model for the study of differentiated cell division. When these bacteria split, they give birth to two slightly different cells, one with a stem and the other with a flagellum.

of five people. *Clostridium tetani* is another sporeformer. Its spores can remain dormant in the soil for years, but when they enter the anaerobic environment of an open wound, they can reactivate and cause tetanus.

Because spores are highly resistant to adverse conditions and can spread easily, it is hard to get rid of them; spores are therefore very dangerous. Take

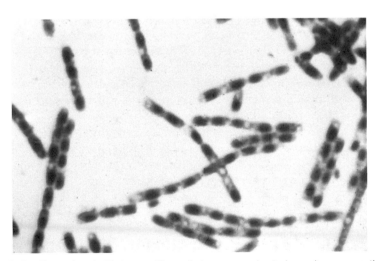

Figure 4. *Bacillus anthracis.* Under conditions of stress, some bacteria produce spores that contain the bacteria's complete DNA. Spores can survive indefinitely in nature until more favorable conditions trigger them to germinate and replicate normally again.

for example the bacterium *Clostridium difficile.* "*C. diff,*" as it is commonly called, is part of the human intestinal microbiome and is highly resistant to most antibiotics. When a patient is treated with antibiotics, the normal intestinal microbiome changes, leaving the resistant *C. difficile* to dominate and cause severe colitis and diarrhea. These bacteria are capable of producing spores that can survive almost anywhere for years. Hence, they are becoming a more and more common cause of health care-associated infections, particularly in hospitals.

Bacteria also have lesser-known survival strategies. For example, some bacteria can halt their peptidoglycan synthesis in order to produce progeny that lack peptidoglycan and are not recognized by the immune system. These are called *L-form bacteria,* from the name of the English surgeon Joseph Lister. Like the mycoplasmas described earlier, they are resistant to many antibiotics and can survive in an infected host for a long time, even during treatment.

It is now possible to watch bacteria divide and to examine the location, the behavior, or the fate of some bacterial proteins. Indeed, new imaging technologies—particularly time-lapse microscopy and superresolution microscopy, both of which use various fluorescent markers—have made it possible to study bacteria in real time. One can observe the precise location of fluorescently linked bacterial proteins (such as the pole or site of division) and see whether they become more intense or disappear during bacterial growth. Combining these imaging techniques with microfluidics—the study of the flow of microquantities of liquids—allows for the real-time observation of bacterial behavior, for example, during changes in cultures or temperature.

Bacterial cell biology is a new discipline that will allow the understanding of bacterial physiology in previously unobtainable detail. It will undoubtedly result in understanding important issues such as the persistence of pathogenic bacteria or the proliferation of some bacteria in certain environments.

CHAPTER 3

The RNA Revolution

Genes of a bacterium—its genetic ID card that distinguishes it from other bacteria—are, like our own, carried by the DNA of its chromosome. Bacterial chromosomes are usually circular in shape. Generally, bacteria have a single chromosome, though some bacteria such as *Vibrio cholerae* have two, and other uncommon genera can have more. *Borrelia*, which is carried by ticks and causes Lyme disease, has many linear chromosomes.

Many bacteria, in addition to their main chromosomes, also have circular minichromosomes called *plasmids*. These chromosomes are not essential to bacterial multiplication but nevertheless may play a significant role in bacterial survival and pathogenicity.

The DNA of a chromosome or a plasmid is a two-stranded polymer. Each strand is a succession of nearly identical components called *nucleotides*, made of a base plus a sugar, that differ only in their base: A, T, G, or C (adenine, thymine, guanine, and cytosine). The two strands of the DNA twist around each other in a helical ladder owing to the affinity of A for T and of G for C. The genes situated

15

along the chromosome are made up of hundreds, sometimes thousands, of nucleotides. They carry within them the information needed to synthesize proteins. In other words, these genes *encode* proteins. In between the genes are found "intergenic" sequences of nucleotides that do not encode proteins.

The DNA present in bacteria is either identically copied or "read." In the first case, the process of *replication*, both strands of the DNA are duplicated exactly, to be passed on to the daughter cells during cell division. In the second case, during *transcription*, the information carried on one DNA strand is "read" by a mechanism that creates a similar but different molecule called a transcript RNA molecule, a *messenger RNA* (mRNA) (Fig. 5). It is called "messenger" because it carries a message from the chromosome that will allow the cell to make a protein. RNAs typically have only one strand of *ribonucleotides* composed of one base and one sugar, as in DNA, but the four bases in RNA are A, U (for uracil), G, and C. RNAs contain the sugar ribose (which gives RNA its name, ribonucleic acid), whereas DNA has deoxyribose (which likewise gives DNA its name, deoxyribonucleic acid).

Replication begins in the part of a chromosome called the *origin of replication* and moves in two directions along the chromosome. As the chromosome is composed of two strands of DNA, each is thus duplicated. Once replication begins, the whole chromosome is replicated. Transcription, in contrast, is a process that moves in one direction only. It can begin at any

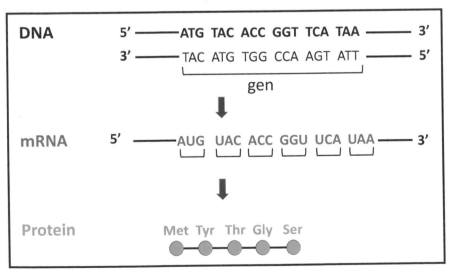

Figure 5. Schematic representation of a double strand of DNA, of its transcribed messenger RNA (mRNA), and of the small protein encoded by the mRNA.

point in a chromosome but only "upstream" of the gene(s) being transcribed, in regions that François Jacob and Jacques Monod named *promoters*. Only certain regions of each DNA strand are transcribed onto RNA. The RNA transcript—the mRNA—is then read during translation.

Translation is a fairly sophisticated process that is achieved by several bacterial actors and in particular a big machine called the ribosome. Ribosomes read the nucleotide sequence of the mRNA transcript by recognizing successive triplets of nucleotides or *codons*. Each of the 64 possible triplets of nucleotides (AUG, UAC, or ACC, for example) corresponds to one of the 20 amino acids that are the building blocks for proteins; thus each amino acid can be encoded by more than one codon. This genetic code is the same for all bacteria and most other organisms. Thus, from their DNA, bacteria produce RNA that is ultimately translated into proteins by this universal code. Bacteria produce thousands of mRNAs, each of which produce proteins.

The results that led François Jacob, André Lwoff, and Jacques Monod to their 1965 Nobel Prize in Physiology or Medicine concerned the discovery that several successive genes, generally involved in the same physiological function, are transcribed together from a single promoter in a single mRNA and thus form a cluster of coregulated genes that they termed an *operon*.

Transcription is not a permanent process. It depends on many environmental factors (such as pH or temperature). It also depends on bacterial factors. In the simplest situation, transcription is regulated by a single "regulator" protein encoded by a gene situated elsewhere on the chromosome. This regulator interacts with the DNA region located upstream of the first gene of the operon; it can either prevent ("repress") or stimulate the expression of mRNA. This model of regulation was studied by François Jacob and his colleagues in the lactose (*lac*) operon, which is involved in the utilization of the sugar lactose, in *Escherichia coli*. The *lac* operon in *E. coli* encodes, among other things, the lactose-metabolizing protein LacZ. Transcription of the *lac* operon is generally repressed, or silenced, because the protein LacI (also called the Lac repressor) binds to the chromosome upstream of the *lac* operon genes and prevents that DNA from being read. When lactose is present in the bacterium's environment, the organism can convert it into allolactose, a molecule that binds to and alters the form of the repressor LacI protein (a conformation, or *allosteric* change) and prevents it from binding to the DNA. In the absence of LacI repression, the *lac* operon genes are transcribed and then translated, allowing the organism to metabolize lactose.

Many researchers have forgotten that in the original model for the operon published in the *Journal of Molecular Biology* in 1961, Jacob and Monod proposed that the repressor is the product of a regulatory gene—that is, an

Repressors and activators

The operon model has been shown to occur in all bacteria, other prokaryotes, and some eukaryotes (particularly in worms such as the nematode *Caenorhabditis elegans*). It has thus been refined and enriched. We now know that genes can be repressed by a variety of repressors that more or less resemble the Lac repressor. Certain genes can be repressed by several repressors but others may also be activated. In this case, it is not a negative regulation that occurs, as for the *lac* operon, but a positive regulation, with an activator protein that binds upstream of the operon when specific conditions require genes to be expressed. One of the best-known activator proteins in bacteria is the protein CAP or CRP in *Escherichia coli*, which binds *cyclic AMP*, a molecule that acts as a hormone and can activate genes involved in the use of sugars other than glucose in bacteria.

Some repressors and activators present in bacteria can also be found in *bacteriophages*, viruses that attack bacteria. Bacteriophages, or *phages* for short, are essentially made of DNA and a few proteins packaged within an outer coat of proteins. The protein cI is a major player in the life of the phage lambda.

Phages, such as lambda phage, can be lytic or lysogenic. A phage is *lytic* if it causes lysis (from the Greek *lysein*, meaning to dissolve) of the infected host cell. Lysis occurs when a phage injects its DNA into a bacterium. The DNA circularizes and proceeds to replicate into new bacteriophages until the sheer number of them makes the bacterium explode. In contrast, a phage is *lysogenic* if the phage's DNA, injected into the bacterium, becomes integrated into the host bacterial cell's chromosome and becomes silent. In some conditions, it can excise and again act as a lytic phage.

The cI protein is a regulator at the heart of both the lytic and lysogenic situations in the lambda phage. cI is able to behave like a repressor when the phage DNA integrates itself into the bacterial genome. cI inhibits the expression of bacterial excision genes that allow the phage DNA to excise from the bacterial chromosome. If bacteria are stressed, irradiated, or grown in certain nutrient deficiencies, the bacterial protein RecA is activated. It cleaves the cI protein, allowing excision to occur.

RNA repressor that would act either in a region of the DNA upstream of the operon, to repress its transcription, or at the start of the operon mRNA, to prevent translation of the message (Fig. 6). Following the publication of this model, researchers analyzed in detail the regulation of the lactose operon and found that the Lac repressor is not in fact an RNA molecule but instead a protein that works upstream of operon genes, as discussed earlier. For years, this discovery was followed by numerous successful discoveries of

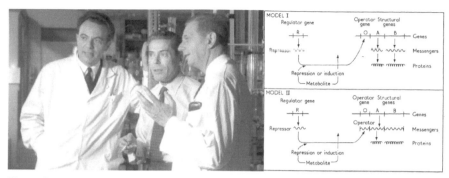

Figure 6. (Left) The three 1965 Nobel Prize winners François Jacob, Jacques Monod, and André Lwoff. (Right) Jacob, Monod, and Lwoff's famous model proposing two possible alternatives for a simultaneous repression of the three genes in the lactose operon. Note that RNA here plays the role of the repressor.

other repressors, forgetting a key part of Jacob and Monod's hypothesis—repressors could also be RNA.

However, in the beginning of the 1980s, it was found that the replication of plasmids—small circular minichromosomes that carry certain accessory genes, such as those for virulence or antibiotic resistance—is controlled by small RNAs that attach (or hybridize) to single strands of the plasmid's DNA and prevent the plasmid's replication. The concept of *antisense* RNA was thus born and the RNA revolution began.

Several other antisense RNAs were then discovered in bacteria, but no one could have foreseen the explosion of knowledge in the early 2000s following the discovery of *microRNAs* in eukaryotes. MicroRNAs are small RNAs with 22 nucleotides that attach to the 3´ region of eukaryotic RNA and affect its translation. This surge of new discoveries was made possible by revolutionary new technologies, in particular the DNA chips called *tiling arrays* and new ultrarapid sequencing methods for genomes and RNA. These techniques allowed for the analysis of the ensemble of RNA transcripts of bacteria grown under a variety of conditions and led to the discovery that bacteria express a large number of RNA transcripts that do not encode proteins. These transcripts, called *noncoding RNAs,* are often the product of the intergenic regions of DNA and can act as regulators. Bacteria can have as many as several hundred distinct noncoding RNAs.

Many RNAs originally categorized as noncoding RNAs or as regulatory RNAs do regulate gene expression themselves. They can also code for small peptides. This is the case for RNAIII of *Staphylococcus aureus*.

While investigating the noncoding RNA in the bacterium *Streptococcus pyogenes*, the research group led by Emmanuelle Charpentier in 2010 made an extremely important discovery. They demonstrated that a small RNA known

Noncoding RNAs

Noncoding RNAs vary widely in size, ranging from tens to hundreds or even thousands of nucleotides. They can interact quite efficiently with other RNAs or even with DNA, as well as with some proteins. They can sometimes act as antisense RNA, even if the noncoding RNA and its target are not entirely complementary. They often prevent mRNA translation by attaching themselves to the translation initiation site. But they can also stimulate translation by attaching to the region of the RNA situated upstream of the genes and changing its structure and configuration in order to stimulate gene translation—for example, by uncoiling a formerly hidden region that prevents ribosomes from accessing their site of action. Noncoding RNA can also bind to proteins, sequester them, and thus prevent them from acting. This situation is thought to be uncommon in nature, although a few examples are well documented, for instance, the CsrA protein that is sequestered by small CsrB RNA in *E. coli*.

Riboswitches: molecular interrupters

Some noncoding RNA elements, called *riboswitches*, function as interrupters. Situated at the beginning of certain mRNAs, a riboswitch can fold in two different ways depending on its binding to its specific ligand. If the riboswitch binds to a ligand, its RNA can take on a form that either impedes the translation of the mRNA (*translational riboswitch*)—in which case the entire mRNA is synthesized but its message is silenced—or stops the transcription of the genes downstream of it (*transcriptional riboswitch*), which leads to the synthesis of a very short RNA. If the riboswitch does not bind to the ligand, the RNA is transcribed in its entirety and is also completely translated. These riboswitch ligands vary greatly in nature, from S-adenosylmethionine (SAM) to vitamin B_1 or B_{12} to transfer RNA or metals such as magnesium.

Riboswitches not only regulate mRNAs as explained above, they can also regulate noncoding RNA. There is such a case in the foodborne pathogen *Listeria monocytogenes*, in which a vitamin B_{12} riboswitch controls an antisense noncoding RNA for the regulator of a series of genes. These genes encode enzymes that process propanediol, a compound present in the intestine that is produced by the fermentation of certain sugars by commensal bacteria. These enzymes require vitamin B_{12} to function as follows: (i) in the presence of B_{12}, the riboswitch leads to the synthesis of a short RNA while allowing synthesis of the regulator protein PocR, and PocR then activates the synthesis of genes under its control; (ii) in the absence of vitamin B_{12}, the riboswitch is configured such that a long form of antisense noncoding RNA is produced, which hybridizes to the mRNA that codes for PocR, stopping its production. Thus the PocR

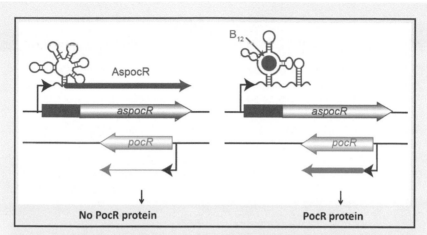

Figure 7. Schematic representation of the chromosome region encoding PocR. In the absence of vitamin B_{12} (left), the long transcript AspocR hybridizes with the transcript for *pocR*, which is then destroyed, preventing the synthesis of PocR. In the presence of B_{12}, the *pocR* messenger RNA allows synthesis of the protein PocR.

activator is not produced unless conditions are favorable, that is, if proteins encoded by the genes that it regulates can be activated by vitamin B_{12}.

Another example of a nonclassical riboswitch is a different vitamin B_{12} riboswitch found in *L. monocytogenes* and *Enterococcus faecalis*, both of which cause intestinal infections. This riboswitch controls a noncoding RNA that can sequester a regulator protein that activates the *eut* genes. *eut* genes code for proteins involved in the utilization of ethanolamine, a compound found in abundance in the intestine. The riboswitch works as follows: (i) if vitamin B_{12} is present, a short form of RNA is produced, a form that does not sequester the regulator protein, which is then free to activate the expression of the *eut* genes; (ii) if vitamin B_{12} is absent, a long form of noncoding RNA is produced that sequesters the regulator protein, which is thus unable to activate the *eut* genes.

This complex alternative process is crucial to the survival of pathogenic intestinal bacteria. Pathogens can use ethanolamine, but only when vitamin B_{12} is present. Since *eut* genes are not present in commensal bacteria, they provide pathogens a significant advantage over commensal bacteria.

as *tracrRNA*, or trans-activating CRISPR RNA, plays a critical role in how the CRISPR system (for clustered regularly interspaced short palindromic repeats) recognizes invading phages and, more specifically, is involved in the destruction of these phages. This research has led to spectacular and unexpected developments, in particular the revolutionary technology called the CRISPR/Cas9 technology for the modification of genomes, as described in the next chapter.

RNAIII in *Staphylococcus aureus*

The RNAIII of *S. aureus* is regulated by *quorum sensing*, which means it is expressed once the bacterial population reaches a certain threshold. RNAIII controls the expression of a certain number of virulence factors. It impedes the translation of proteins, such as protein A, expressed on the surface of the bacterium or secreted during the beginning of an infection. It also impedes the translation of transcription regulators such as RotA. However, it activates the expression of the toxin known as alpha-hemolysin (Hla) by acting as an antisense that allows the corresponding RNA to be translated. Additionally, RNAIII codes for the small Hld protein, another toxin of 26 amino acids. The 514-nucleotide-long RNAIII in *S. aureus* is a very active molecule and thus can regulate many facets of bacterial physiology over the course of an infection.

The excludon

Some RNAs function both as antisense and as messenger. They are encoded in recently discovered regions of bacterial chromosomes called *excludons*. These regions were originally detected in the *Listeria* genome but were then found to exist in various other bacteria. Excludons are made up of two DNA regions encoding genes or operons that are oriented in opposition to one another on the bacterial chromosome. They encode a long RNA (up to 6,000 nucleotides) that is antisense to one of the regions. The first part of this RNA functions as an actual antisense that has a negative effect on the genetic expression of the gene or operon located on the strand opposite to the one that codes the RNA. But the second part of the RNA can act as an mRNA (Fig. 8).

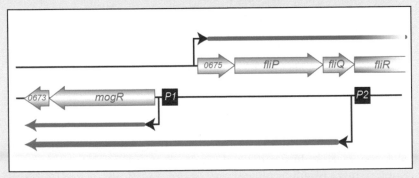

Figure 8. Example of an excludon. Once the transcription beginning at P2 is expressed and generates a long transcript, the operon on the right becomes less expressed.

CHAPTER 4

From the CRISPR Defense System to the CRISPR/Cas9 Method for Modifying Genomes

In nature, bacteria need to defend themselves constantly, particularly against bacteriophages (or phages), the viruses that specifically attack bacteria. A phage generally attaches itself to a bacterium, injects its DNA into it, and subverts the bacterium's mechanisms of replication, transcription, and translation in order to replicate itself. The phage DNA reproduces its own DNA, transcribes it into RNA, and produces phage proteins that accumulate to generate new phages and eventually cause the bacterial cell to explode (or lyse), releasing hundreds of new bacteriophages. Phages continually infect bacteria everywhere—in soil, in water, and even in our own intestinal microbiota (Fig. 9). Bacteriophage families are numerous and vary widely in their form, size, composition, and the bacteria they target.

To begin their attack, bacteriophages need a site of attachment, a particular component on the surface of a bacterium. This site of attachment is specific for each virus and the bacteria that it can infect.

Infections of bacteria by phages are of great concern, particularly in

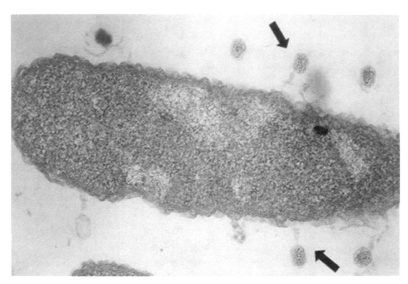

Figure 9. Bacteriophages infecting an *Escherichia coli* bacterium.

the dairy industry, which uses certain bacteria, for example *Streptococcus thermophilus*, to make yogurt and cheese. *S. thermophilus* transforms the lactose in milk into lactic acid. Additionally, each different bacterial strain contributes its own unique taste and texture to the yogurt, which must remain consistent to ensure a reliable product and successful sales. If a bacterial strain disappears as the result of a bacteriophage infection, the economic consequences for the manufacturer can be disastrous.

One of the great discoveries of this decade is that bacteria have an immune system called *CRISPR*, for clustered regularly interspaced short palindromic repeats, meaning small, regularly spaced, palindromic (the sequence reads the same from either end) repeats. CRISPR regions in the chromosomes allow bacteria to recognize predators, particularly previously encountered phages, and to destroy them. CRISPR regions protect and essentially "vaccinate" bacteria against bacteriophages.

In fact, it has been shown that bacteria can be artificially vaccinated! When a population of bacteria is inoculated with a phage, a small number survive and are able to integrate a fragment of the phage DNA into their genome, in the region called the CRISPR locus. This allows the bacteria, if the phage ever attacks again, to recognize the phage DNA and degrade it. This ingenious phenomenon, known as *interference*, occurs due to the structure of the CRISPR region and to *cas* genes (*CRISPR-associated* genes) located near this region.

The CRISPR locus is a region of the chromosome composed of repeated sequences of around 50 nucleotides, interspersed with sequences known as *spacers* that are similar to those of bacteriophages. Some bacteria have several CRISPR loci with different sequence repetitions. Around 40% of bacteria have one or more CRISPRs, whereas others have none. CRISPR loci can be quite long, sometimes with more than 100 repetitions and spacers. CRISPRs have two functions: acquisition and interference. *Acquisition*, also called *adaptation*, is the process of acquiring fragments of DNA from a phage, and *interference* is the immunization process by Cas proteins encoded by *cas* genes (Fig. 10).

Bacteria have numerous proteins with various complementary and synergistic functions in the process of adaptation and interference. They permit the addition of DNA fragments into the CRISPR locus, but their main purpose is to react to invading phages. The CRISPR locus is transcribed into a long CRISPR RNA, which is then split into small RNAs called crRNAs, each containing a spacer and a part of the repeated sequence. When a phage injects its DNA into the bacterium, the crRNA recognizes and binds to it. An enzyme then recognizes the hybrid and cleaves the phage DNA at the point where the crRNA has paired. Replication of the phage DNA is inactivated, and the infection is stopped.

One of the key discoveries that led to the use of CRISPR systems in what is called "genome editing," or *modification*, was the identification of the proteins involved in the cleavage of the hybrid DNA. This process is performed by a complex of proteins containing the protein Cas1 and sometimes by a single protein called Cas9. Cas9 is unique in that it can attach itself to a DNA strand and, due to the two distinct domains of its structure, cut this DNA on each of its two strands. This protein is the basis of the CRISPR/Cas9 technology, which enables a variety of genome modifications and mutations in mammals, plants, insects, and fish in addition to bacteria. This system works due to the Cas9 protein and also a guide RNA hybrid that is made from one RNA similar to the region to be mutated and a second RNA called *tracrRNA*, or trans-activating crRNA. tracrRNA was discovered next to the CRISPR locus in *Streptococcus pyogenes* and was shown to be homologous to the repeated regions of the locus, enabling it to guide the Cas9 protein and the crRNA toward the target.

In summary, by expressing the Cas9 protein with a composite RNA made up of an identical sequence to the target region, a tracrRNA, and a complementary fragment to the tracrRNA, one can now introduce a mutation or deletion into a target genome of any origin.

After the 2012 publication in *Science* of the elegant studies by the teams led by Emmanuelle Charpentier and Jennifer Doudna, the CRISPR method was so intriguing that it provoked an avalanche of research and publications

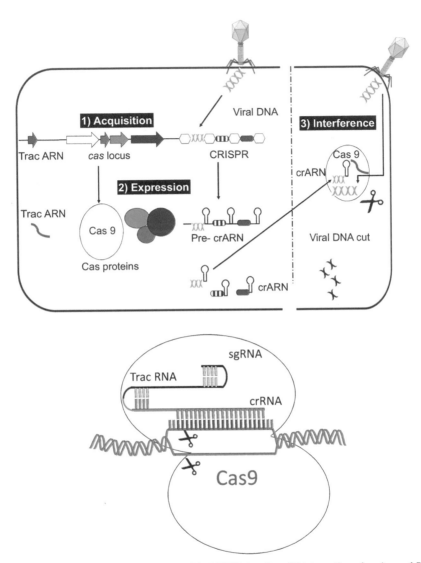

Figure 10. (Top) The three steps involved in CRISPR function. (1) Integration of a piece of DNA from a phage into the CRISPR locus (acquisition); (2) the expression of Cas proteins and of pre-crRNA, which is then split into small crRNAs; (3) the interference that takes place when the DNA injected by a phage into a bacterium meets a crRNA, a hybrid form that is then degraded, consequently preventing infection. (Bottom) Schematic drawing of genome modification (gene editing) by an sgRNA (small guided RNA) made of crRNA and tracrRNA and the endonuclease Cas9.

demonstrating that this technique could be used in many cases and with many variations. For example, a Cas9 protein named *dCas* (dead Cas), if fused to a repressor or activator protein, can attach to the desired locus without cleaving it, then activate or repress genes in mammals. It was also shown that the CRISPR method can generate multiple mutations at a time if performed with a single Cas9 but with a variety of guide RNAs.

It is thus through the studies of microbiologists interested in fundamental aspects of bacterial physiology—such as resistance to phages, the role of the noncoding repeated sequences found in many genomes, and the role of small noncoding RNAs—that a revolutionary technique was born. CRISPR has revolutionized many domains of biology, to the point that medical applications for this technique are now within reach, such as targeted gene therapy. The researchers most involved in these discoveries have already been recognized for their contributions by undoubtedly well-deserved awards.

While being the object of intense study, the CRISPR/Cas9 technology raises important ethical questions. Should we start using it now for gene therapy? Do we have enough experience and perspective on the matter to make this decision? How can we be sure we are not creating unintentional mutations along with the targeted mutations? Should we have ethical concerns even with the latest technical developments and the use of modified Cas9 proteins? These issues are the focus of important international ethics committees.

CHAPTER 5

Antibiotic Resistance

The discovery of antibiotics

Sir Alexander Fleming, studying the properties of *Staphylococcus* bacteria at St. Mary's Hospital in London in the late 1920s, returned from vacation to find that he had accidentally left out a number of culture plates, one of which had become contaminated by mold. At that time, bacterial cultures were made in liquid or agar media in Petri dishes. In a dish of growing staphylococci, Fleming noted a clear area without bacterial growth around the moldy patch, while the rest of the bacteria on the plate had grown normally. From this, he deduced that the fungus might be capable of actually killing the bacteria. Fleming rapidly identified the invader as *Penicillium notatum* (now known as *Penicillium chrysogenum*). He created an extract from this fungus that he found was effective not only against staphylococcus but also against the bacteria responsible for scarlet fever (group A *Streptococcus*), diphtheria (*Corynebacterium diphtheriae*), pneumonia (pneumococcus), and meningitis. He named the active agent in his fungus extract *penicillin*. Fleming published his discovery in 1929, noting in passing that penicillin could be useful injected or applied as an effective antiseptic agent against bacteria. Unfortunately, penicillin was initially very difficult to purify and isolate in quantities sufficient for testing. It was not until a decade later that Australian pharmacologist Howard Florey and German biochemist Ernst Chain successfully isolated penicillin. Trials of penicillin in humans that began in 1941 had spectacular results. In May 1943, British soldiers fighting in Algeria received the first injections of penicillin, produced in the United States. Fleming, Florey, and Chain shared the 1945 Nobel Prize in Physiology or Medicine for their work on penicillin and its therapeutic applications.

Following Florey and Chain's work, other researchers, notably Selman Waksman, investigated thousands of microorganisms in a hunt for other antibiotics that, unlike penicillin, did not kill all species of bacteria. In 1944, streptomycin was discovered in a species of *Streptomyces* bacteria found in soil. Since then, researchers have isolated thousands more antibiotics.

Sulfonamides enter the game

Even before penicillin was used in humans, other antimicrobial compounds were being investigated. Early in the 20th century, attention was drawn to the unexpected antimicrobial properties of the diazoic chemical dyes such as Prontosil. German biochemist Gerhard Domagk is credited with discovering the active form of the drug sulfonamidochrysoidine, which was further investigated in 1935 by Jacques and Thérèse Tréfouël, Frédéric Nitti, and Daniel Bovet at the Pasteur Institute in Paris. Sulfonilamide, a sulfonamide antibiotic, was used for more than 30 years to treat streptococcal infections, particularly the skin infection called erysipelas. Sulfonamides are derived from *para*-aminobenzoic acid and block the synthesis of tetrahydrofolate, thereby inhibiting the synthesis of purines and pyrimidines, the bases that form the DNA, leading to bacterial death. Unfortunately, sulfonamides can cause dramatic side effects such as allergic reactions.

Gerhard Domagk was awarded a Nobel Prize in Physiology or Medicine in 1939 for his discovery of sulfonamidochrysoidine, although he was not allowed to accept it until after World War II.

Mode of action of antibiotics

In general, antibiotics are chemical substances that specifically target bacteria. Because of this specificity, antibiotics differ from antiseptics, which are only used externally. Antibiotics can prevent bacterial growth (*bacteriostatic* action) or completely destroy the bacteria (*bactericidal* action). There are now more than 10,000 known antibiotic molecules, around 100 of which are used in medicine.

Antibiotics are either bactericidal or bacteriostatic depending on their target. Penicillin and other antibiotics in the beta-lactam family prevent the synthesis of the cell wall. Some antibiotics, including cyclic peptides such as polymyxin B, target and modify the cell membrane, leading to the leaking of bacterial products and cell death. Other drugs, such as fluoroquinolones, enter the cell and attach to bacterial DNA, preventing replication and transcription. Some antibiotics, like the sulfonamides described above, are structurally analogous to bacterial compounds involved in DNA synthesis; they compete with and replace these compounds and thus block replication. Tetracyclines (Aureomycin, or chlortetracycline) and macrolides (erythromycin) act during specific steps in the synthesis of bacterial proteins to prevent bacterial growth.

The first antibiotics were produced naturally from bacteria such as *Streptomyces* or fungi such as *Penicillium*. Many antibiotics currently used are made from compounds isolated in natural products and then modified. They are therefore referred to as *semisynthetic antibiotics*, while others are completely synthetic.

Antibiotics used in medical therapy are specifically chosen because they target the infecting bacteria but not human cells. However, they are not entirely harmless and can have side effects, especially when used over a long period of time or in large doses. Most notably, antibiotic use alters the intestinal microbiome and can result in colitis and diarrhea. Penicillins, cephalosporins, sulfa drugs, and others can cause allergic reactions. And certain antibiotics are toxic to human tissues and can have serious side effects. Gentamicin can cause hearing loss or kidney failure, streptomycin can also cause hearing loss, and fluoroquinolones may cause heart trouble. Most side effects of antibiotics resolve once the antibiotic is stopped, but not always.

Each type of antibiotic is generally active only against specific bacteria or families of bacteria (Fig. 11). Antibiotics became widely used after World War II and have considerably reduced the mortality rate of infections, including tuberculosis and the plague. Unfortunately, their heavy and widespread use in human and animal health has led to the development of strains of antibiotic-resistant bacteria. First noticed toward the end of the 1960s, this

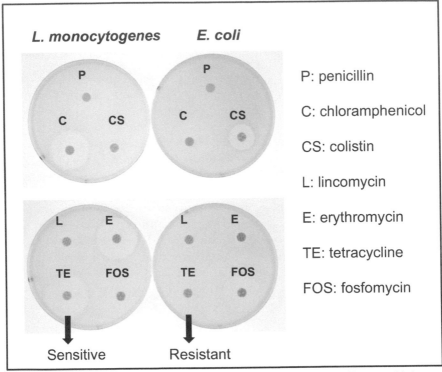

Figure 11. An antibiogram is an antibiotic sensitivity test. Disks containing different antibiotics are placed on a layer of bacteria grown in a petri dish. A clear zone around the disk can be seen when bacteria susceptible to the antibiotic are unable to grow or have been killed. If bacteria are resistant to the antibiotic, they will be able to grow around the antibiotic-containing disks.

phenomenon has grown in scope and scale to the point of becoming a major global concern. New regulations and research for alternative solutions are needed to keep us from coming to a therapeutic dead end.

Antibiotics in animals

About 50% of all antibiotics are used in livestock to treat disease, to prevent disease from spreading, and to enhance healthy growth. According to a report published by the French National Agency for Veterinary Medicinal Products (ANMV), France was the second largest consumer of antibiotics in Europe in 2009; 44% of all antibiotics sold in France that year were for raising pigs, 22% for raising poultry, and 16% for cattle. The use of antibiotics to enhance livestock growth has been prohibited in Europe since 2006;

however, in the United States, antibiotics are still systematically used for this purpose. As reported in 2015, 100,000 tons of antibiotics were consumed in the United States, India, and China in 2010, with nearly two-thirds used in livestock. As in humans, the overuse of antibiotics in animals can cause increasing bacterial resistance to one or more antibiotics (*multidrug-resistant, or MDR, bacteria*). If this occurs in domesticated animals raised for human consumption, the resistant bacteria can spread to humans.

From the first cases of resistance to a global panic

Certain bacteria are naturally resistant to certain antibiotics due to their impermeable cell walls or the lack of a specific antibiotic target. For example, *Escherichia coli* is naturally resistant to vancomycin, *Pseudomonas aeruginosa* is resistant to ampicillin, and *Listeria* strains are resistant to nalidixic acid.

However, the problem now is that many pathogenic bacteria that were initially susceptible to one or more antibiotics have acquired or developed resistance genes. These genes encode proteins involved in a variety of mechanisms, such as the ability to modify or destroy the antibiotic, to modify the antibiotic's target in the bacterial cell, to make the cell's membrane impermeable to the drug, or to develop proteins that pump the antibiotic out of the cell.

Genetic resistance can be acquired by several means. First, the bacterium may simply mutate. This can occur during replication and lead to antibiotic resistance. The mutation(s) allows the mutant bacterium to survive and reproduce in the presence of an antibiotic while its unmutated (*wild-type*) siblings cannot, and it can then disseminate into the environment.

The other scenario is the acquisition via *horizontal gene transfer* of resistance genes present on a plasmid in another bacterium. This occurs via the phenomenon known as *conjugation*, when plasmid DNA containing resistance genes is transferred directly from one bacterium to another through a tube called a pilus. Roughly 80% of acquired resistance results from plasmid transfer during bacterial conjugation.

Horizontal gene transfer can also occur via *transformation*. Under certain circumstances, some bacteria become "competent," meaning they can accept external DNA from their environment, generally from bacteria that have lysed. This type of gene transfer has been well documented in *Streptococcus pneumoniae*, and there are more than 40 other species of bacteria known to acquire external DNA by natural transformation.

Resistance to antibiotics frequently appears in hospitals. In France, for example, hospitals are where half of human antibiotics consumption takes place. Some of the most serious problems with bacterial resistance are linked to methicillin-resistant *Staphylococcus aureus* (MRSA), which is responsible for various lung, bone, or blood infections, particularly in patients in intensive care. Infections with resistant strains last longer and have a higher mortality rate; it is estimated that patients with MRSA have a 63% higher mortality rate than those infected by nonresistant *S. aureus*. Antibiotic resistance increases the cost of health care due to the intensive care and prolonged hospital stays required.

In hospitals, *P. aeruginosa* is responsible for many nosocomial, or *health care-associated*, infections resistant to antibiotics in the carbapenems family. In particular, *P. aeruginosa* is a significant cause of morbidity and mortality in cystic fibrosis patients. *Acinetobacter baumannii* strains, which can cause pneumonia and meningitis, have become more and more prevalent in health care-associated infections because they are naturally competent and have become resistant to various antibiotics.

Resistant bacteria encountered outside the hospital often include penicillin-resistant pneumococci and enterobacteria such as *E. coli* and *Klebsiella pneumoniae* that produce broad-spectrum beta-lactamases, meaning they are resistant to different types of beta-lactams. *E. coli* bacteria that cause urinary infections have become resistant to amoxicillin and must now be treated with cephalosporins.

Tuberculosis, caused by *Mycobacterium tuberculosis*, was first treated with streptomycin. Over the past few decades, tuberculosis therapy has evolved into a combination of four different antimicrobials (isoniazid, rifampin, pyrazinamide, and ethambutol) with a treatment course of 6 months. This extended period is necessary due to the slow growth of *M. tuberculosis*, which multiplies inside the patient's macrophages and forms compact structures of aggregated infected cells called *granulomas* that are difficult for drugs to reach. Tuberculosis resistant to the first-line drugs isoniazid and rifampin is emerging, often due to an inadequate treatment period or the use of inappropriate or low-quality antituberculosis drugs. Second-line drugs are available, but they are expensive and scarce, may need to be taken for as long as 2 years, and can have serious side effects. Furthermore, MDR and extremely drug-resistant (XDR) strains of *M. tuberculosis* have been observed, especially among immunosuppressed patients and those in developing countries.

Potential solutions and hopes

After World War II, antibiotics rapidly transformed medicine, curing bacterial infections that were previously often fatal. Antibiotics also allowed for spectacular progress in survival after major surgery, organ transplant, or chemotherapy and the treatment of infections in patients with immune deficiency. However, antibiotic resistance has now led to an increase in health care-associated and other infections that are difficult or even impossible to treat. Will we return to a preantibiotic era?

The golden age of antibiotics began to wane in the early 1990s as the research and development of new antibiotics slowed down. Bacterial resistance to antibiotics was not yet widespread, but scientists were beginning to notice increasing numbers of antibiotic-resistant bacteria and to foresee the inevitable consequences. In 2002, the French national health insurance launched a campaign, "*Les antibiotiques, c'est pas automatique*" (Don't automatically go for antibiotics), whose aim was to inform patients that antibiotics should only be used in cases of bacterial infection and never for viral infections. As a result, the consumption of antibiotics in France fell by 15%. A second campaign to broaden the public's understanding about resistance was begun in 2010, using the slogan "*Si on les utilise à tort, ils deviendront moins forts*" (If you use them wrong, they become less strong). Today, in health care institutions in France and around the world, utmost importance is placed on limiting the use of antibiotics and fighting the spread of resistant bacteria by focusing on strict hygiene.

Can we discover new antibiotics? This requires extensive research and development that unfortunately is not a high priority for pharmaceutical companies. As mentioned above, the best antibiotics were found in bacteria or other microorganisms in soil. It seemed that these microorganisms had been mostly exhausted as a source of new antibiotics. However, the environment is full of bacteria, both culturable and (thus far) unculturable. A completely new antibiotic, teixobactin, was recently discovered; it is produced by a previously unknown Gram-negative bacterium, *Eleftheria terrae*, which belongs to a new genus similar to *Aquabacterium*. *E. terrae* was originally considered unculturable but was found to grow in a medium similar to soil. Teixobactin is quite effective against Gram-positive bacteria such as staphylococci, enterococci, and *M. tuberculosis*. It is also active against *Clostridium difficile* and *Bacillus anthracis*, although it is ineffective against most Gram-negative bacteria. Teixobactin inhibits bacteria by attaching to lipid II, a precursor to peptidoglycan synthesis; *in vitro* experiments suggest that bacteria do not easily develop resistance to it. The last powerful new antibiotic to go on the market was vancomycin in 1956. Resistance to vancomycin developed due to gene transfer from strains close to the strain that produced the antibiotic—*Amycolatopsis orientalis*—but only after about 30 years of use. One can hope for a similar situation with teixobactin.

New antibiotic-hunting strategies are being investigated that could improve on the laborious process of sifting through countless bacterial cultures. Researchers are testing the effects of whole libraries of chemical

Inhibition of quorum sensing

Many pathogenic bacteria express their virulence factors only when the bacterial population is at a high density. Receptors located on the cell surface detect molecules, called *autoinducers*, produced by other bacteria. The ability of bacteria to detect a level of autoinducer that is sufficient to trigger the expression of virulence factors is called *quorum sensing*.

Because quorum sensing orchestrates virulence in pathogenic bacteria, inhibiting it should inhibit their virulence. To inhibit quorum sensing, therefore, would involve either deactivating the enzymes that create the signaling molecules, deactivating the cell-surface receptors for those molecules, or otherwise interfering with the signaling mechanism. This last principle has already been proven in the case of cholera. (See more on quorum sensing in Part 2.)

There is a particular form of quorum sensing in which peptides produced by bacteria can act as a weapon against other bacteria that do not produce these proteins, inducing the suicide of the nonproducer bacterium (also see chapter 7). The attacking bacterium produces a toxin, an endonuclease that destroys the mRNA of the targeted strain. Normally the targeted strain produces an antitoxin to protect itself from this, but when the attacker floods the area with peptides as well, this causes environmental stress in the target. The stressed nonproducer stops making its antitoxin, which allows the attacking bacterium's toxin to destroy it. This system of killing bacteria by using peptides is still relatively unknown and its use warrants further research.

compounds on the growth of bacterial cultures. For example, the team led by Stewart Cole discovered that benzothiazinones kill *M. tuberculosis* by blocking the synthesis of arabinan, a compound in mycobacterial cell walls. PBTZ169, a benzothiazinone derivative that is synergetic with bedaquiline and pyrazinamide, is a very promising antibiotic. Another approach involves inhibiting the functions of bacterial proteins encoded by essential genes that do not have equivalents in mammalian cells. Essential genes are those whose products are required for the bacterium's survival and that cannot mutate without killing the bacterium.

Finally, *inhibition of quorum sensing*, described above and in chapter 7, provides fuel for researchers' dreams of halting infections by any means possible.

Phage therapy

Phage therapy is making headlines again. This strategy eliminates bacteria by infecting them with bacteriophages, the viruses that attack bacteria. The advantage of this approach is that particular bacteriophages infect only specific strains of bacteria; they are not known to cause any side effects, and they act rapidly, often producing immediate results. However, since bacteriophages do not enter into other types of cells besides bacteria, they cannot be used to treat intracellular bacteria. They are generally administered topically.

Phage therapy was beginning to be widely used in the early 1900s but mostly disappeared from use following the discovery of antibiotics (except in former Soviet Union countries such as Georgia). However, interest in phage therapy has revived since the discovery that phages could improve the success of skin grafts by reducing *P. aeruginosa* infection. Many recent studies have further validated the strategy. Extensive "libraries" of phages exist at multiple institutions, including the Eliava Institute in Georgia and the Institute of Immunology and Experimental Therapy in Poland.

Phage therapy is not yet legal in France nor in other European countries. To be allowed on the market in France, it will have to wait for the results of studies approved by the French DGA (*Direction Générale de l'armement*) or the European Union. A collaborative project called Phagoburn, founded in 2013 and financed by the European Union Seventh Framework Program of Research and Development, aims to evaluate the results of phage therapy on topical infections caused by *E. coli* and *P. aeruginosa* in burn patients. Now conducted in hospitals treating major burn patients in France (Hôpital d'instruction des armées [IIIA] Percy, in Clamart), Belgium (Military Hospital Queen Astrid, Brussels), and Switzerland (CHU Vaudois, Lausanne), this study is the first of its kind and promises valuable results.

Phages could also be used prophylactically, for example, to treat potentially infected raw food to prevent infection outbreaks. In 2006, the U.S. Food and Drug Administration authorized the use of phages to treat food products and prevent *Listeria* contamination of food; however, no individual phages have yet been authorized for this process.

Bdellovibrio: why not?

Another potential approach is based on using a killer bacterium to kill bacteria. *Bdellovibrio bacteriovorus*, a part of our intestinal microbiota, is a small Gram-negative bacterium that almost exclusively attacks other Gram-negative bacteria, such as *E. coli* or *Acinetobacter baumannii*. *Bdellovibrio* penetrates the host bacterium's outer membrane and establishes residence in the *periplasm*, the space between the outer and inner cell membranes, where it grows and divides, becoming a *bdelloplast*. Ultimately the host cell lyses, releasing a horde of new bdellovibrios into the environment. Researchers are now speculating that this living antimicrobial agent could be used in combination with an antibiotic as a first-line treatment for external use on skin, for example, for burns.

We are living in a time of transition, where it is critical both to prevent the emergence of further antibiotic resistance and to make every effort to develop new and effective therapeutic strategies.

PART II

Sociomicrobiology: The Social Lives of Bacteria

CHAPTER 6

Biofilms: When Bacteria Gather Together

One of the main differences between prokaryotes and eukaryotes, besides the fact that prokaryotes (bacteria and archaea) do not possess a nucleus, is that prokaryotes produce two identical, unicellular daughter cells when they divide. In contrast, eukaryotes (animals and plants) give birth to highly complex, multicellular organisms with differentiated tissues and organs; the cells split but do not necessarily generate identical cells. Each cell of a higher organism contains the same DNA, but not all genes are expressed in all cells. During development, the cells of higher organisms specialize to form tissues and organs.

A certain form of multicellularity, named for the first time as a *biofilm* by J. W. Costerton in 1978, is maybe closer to reality than unicellularity. We are now learning that this may be the natural way of life for nearly all bacteria.

Classic microbiology, as in the days of Louis Pasteur and Robert Koch, generally examined bacteria by growing them in pure culture in rich liquid media, conditions ideal for a mode of growth known as "planktonic" growth. However, these culturing conditions are often far different from the bacteria's natural growth conditions. More and more, scientists are realizing that bacteria can adopt planktonic form or assemble into physiologically distinct biofilms, depending on the conditions they encounter (Fig. 12).

Biofilms form when bacteria encounter and adhere to a surface and then grow together to create a complex community that often forms a special type of structure. Biofilms can be composed of a single species of bacteria or contain multiple species; natural biofilms may also contain fungi and amoebae. In a biofilm, bacteria produce a set of compounds called the *matrix* that maintains the biofilm cohesion, protects it from external effects, stimulates certain properties within its members, and allows synergistic interaction. The bacteria present in biofilms are more resistant to oxygenated water, bleach, and other disinfectants than planktonic bacteria are.

Because biofilms also provide their members with a heightened resistance to antibiotics, they constitute an increasing problem in the medical setting. Not only do biofilms grow on inert surfaces, contaminating whatever comes into contact with these surfaces, including food, but they can also form on the surface of teeth, causing cavities and gingivitis, and develop on prosthetics and other medical devices, such as joint implants or inside catheters, thereby potentially contaminating nutrients delivered to patients.

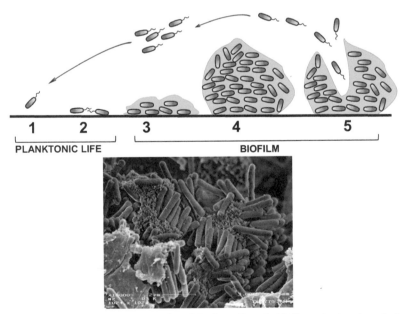

Figure 12. (Top) Schematic representation of the formation of a biofilm and return to a planktonic life. (Bottom) A biofilm image taken with an electron microscope.

Formation and maturation of a biofilm

The formation of biofilms is the object of intensive studies. The process involves two stages: attachment followed by maturation.

A motile bacterium senses a surface by means of its pili, which act as "mechano-sensors"; the pili are capable of inhibiting the rotation of the bacterium's flagella, which in turn stimulates the synthesis and production of extracellular polysaccharides. These polysaccharides form at one pole of the bacterium and essentially irreversibly glue the cell to the surface. This is what happens with *Caulobacter crescentus* and the plant pathogen *Agrobacterium tumefaciens*. This response involves a signaling molecule within the bacterium, cyclic di-GMP, that acts as a hormone.

Another means of stopping bacterial motility has been observed in the soil bacterium *Bacillus subtilis*. One of the first events that follows contact of *B. subtilis* with a surface is the modification of its flagella by the addition of a sugar, which inhibits their capacity to rotate. For these bacteria, a complex system of regulation by temperature, pH, etc., ultimately leads to the formation of a biofilm. Once attached, the bacteria grow and divide while secreting the biofilm's matrix.

Biofilm formation is reversible; however, under some conditions, a biofilm will disperse, returning its component bacteria to independent, planktonic growth.

To all appearances, biofilms are living structures. They are not impermeable, but allow water and other liquids and nutrients to penetrate. Their matrices contain polysaccharides (sugar polymers produced by bacteria). Biofilms often contain DNA from cells that have burst after losing their cell walls. They can even contain cellulose; for example, salmonellae produce cellulose as efficiently as plants do!

Since biofilms are the natural way of life for most bacteria, they cause problems in many human arenas, including medicine (resistance to antibiotics, persistence of contamination), industry (corrosion problems—certain biofilms are involved in the development of mineral deposits), and agriculture. Some bacteria, such as *Listeria*, may seem to disappear after thorough cleaning or decontamination, but they then reappear up to years later after having survived in biofilms, for instance inside milk containers. Biofilms can also pose a threat to the quality of potable water and cause trouble for maintenance of water pipelines.

Biofilms are not the only form of microbial cooperation. Although biofilms require adhesion to an appropriate surface in order to form, there are other microbial groupings much less organized and more flexible in composition. These are called microbiota (see chapter 9).

CHAPTER 7

How Bacteria Communicate: Chemical Language and Quorum Sensing

Whether in biofilms or other microbial assemblies, bacteria communicate. They speak to each other! Their chemical language consists of complex molecules that they release into the environment. At the same time, they gauge the concentration of these signal molecules in their surroundings via sensors on their surface or in their cytoplasm. This allows them to actually measure the number of other bacteria around them! This phenomenon is known as *quorum sensing*.

In nature, a wide range of different bacteria live together and use a variety of signaling molecules to communicate. Because each bacterial species has its own chemical language, not all bacteria can communicate with each other. However, some bacteria are bilingual or multilingual and can respond to different signals in different ways, recognizing identical (sister) bacteria or similar (cousin) bacteria. Some signals are specific to a single species, while other, less specific signals can correspond to a bacterial genus, which contains many species.

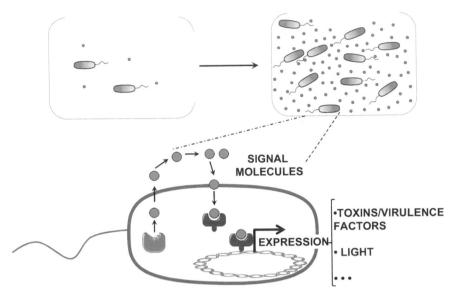

Figure 13. Schematic representation of the quorum-sensing effect. Bacteria release signal molecules into the environment. They are then able to recognize these molecules by means of cell receptors located on their surface or in their internal milieu. This recognition, in turn, triggers the expression of virulence factors, light-generating molecules, or other factors.

What is the point of quorum sensing? It allows bacteria to coordinate their behavior and act together as if they were one multicellular organism (Fig. 13). For example, this is how pathogenic bacteria know to wait to produce virulence factors and provoke infections until their bacterial density is high enough to thwart the host's immediate response, giving the infection the best chance of being successful.

Quorum sensing is used not only by pathogenic bacteria. Some light-producing bacteria function only when they are in groups. In fact, quorum sensing was first discovered in bioluminescent bacteria, notably in a *Vibrio* species living in squid (see discussion in chapter 9).

The signaling chemicals that bacteria use to communicate are not very complex molecules, but there are various types: simple molecules derived from products of the bacterial metabolism, such as homoserine lactone, or slightly less simple molecules that contain rare chemical elements such as boron. In Gram-positive bacteria, signaling molecules are often very small peptides, modified in various ways.

Research on quorum sensing has raised hopes in the fight against pathogenic bacteria, particularly those resistant to antibiotics. It is possible to intervene in the process of quorum sensing, interfering with the ability of the bacteria to recognize each other and to act in groups. Finding a molecule

Interrupting quorum sensing to prevent infection

The *Escherichia coli* strain Nissle, currently used in some treatments for diarrhea, produces a signaling molecule, AI-2. Faping Duan and colleagues introduced the expression of an alternative quorum-sensing molecule, CAI-1, into this *E. coli* strain in order to interfere with the regulation of virulence genes at the molecular level in the cholera bacterium *Vibrio cholerae*.

In mice treated with this strain of *E. coli* and then infected with *V. cholerae*, the survival rate rose by 90% in comparison with nontreated mice. This research highlights the possibility of treating infections by interfering with quorum sensing.

Suicide—or murder?

It is possible for a bacterium of one species to induce a bacterium of another species to commit suicide. The first species releases into the environment a small peptide of five amino acids, such as EcEDF in *E. coli*. A high concentration of this peptide can cause stress for a bacterium of the other species. The second bacterium normally protects itself against a toxin by generating an antitoxin that blocks the action of the toxin. Interaction with the foreign peptides induces stress in the second species, causing degradation of the antitoxin and thus allowing the toxin to act, leading to *programmed cell death*, also called *apoptosis*, or "cell suicide." This mechanism is a source of great interest as a potential alternative to antibiotics.

that prevents *Pseudomonas* bacteria from recognizing each other, for example, would prevent them from becoming pathogens. *Pseudomonas* is the most dangerous infection for patients with cystic fibrosis.

It has been shown that there is a system of quorum sensing that allows members of one species of bacteria, if they exist in a high enough density, to kill members of another species, or more specifically, to force the other to commit suicide (see the above box)! This system is far from being the only one put into play by bacteria to kill off their fellow bacteria. More will be discussed in chapter 8.

Furthermore, it is important to note that bacteria can internalize signaling molecules. These molecules enter by passive transfer or an uptake system,

Gene transfer: conjugation, transformation, and formation of nanotubes

Bacteria can exchange genetic material, including chromosome fragments and plasmids, during *conjugation*. Conjugation occurs when a bacterium generates pili that attach to a second bacterium and DNA is then exchanged from one cell to the other through the pili. This is how resistance genes can spread or how large fragments containing virulence genes known as *pathogenicity islands* are transmitted from one strain to another.

The transfer of genetic material can also occur by *transformation* in bacteria that have become competent—that is, capable of absorbing DNA straight from the environment, usually from lysed bacteria.

It has recently been found that bacteria can form nanotubes, very fine membrane tubes that permit the exchange of compounds. However, these structures remain somewhat controversial, as genes regulating their formation have so far not been identified.

and the receptor, instead of being on the surface, is internal. Besides signaling molecules, bacteria can also acquire genetic material from fellow bacteria by the process called *gene transfer*.

CHAPTER 8

When Bacteria Kill Each Other

In all domains of life, the fight for survival naturally favors the fittest, that is, those best adapted to the surrounding conditions. This is the phenomenon of natural selection. The transmission of acquired characteristics contributes to evolution and the development of new species. Although Charles Darwin (1809–1882) was not familiar with bacteria, his ideas are readily applicable to them. The bacterial world, just like the finches Darwin observed in the Galapagos Islands, is constantly adapting and evolving.

Evolution selects for the best-adapted bacteria. When external agents such as antibiotics or bacteriophages attack bacteria, the strains that survive and thrive are the ones that can best protect themselves by rapidly acquiring resistance against the antibiotics or by "vaccinating" themselves against the phages.

Antibiotics and phages are not the only antibacterial agents. Bacteria can also be attacked and killed by other bacteria. We have mentioned, for example, that the small predator *Bdellovibrio* can invade other bacteria and multiply, causing the host bacteria to explode. However, other more subtle strategies exist. We have discussed how a type of quorum sensing induces the programmed cell death of bacteria that respond to a small peptide released by the "killer" bacteria. In that situation, the targeted organism is stressed by the peptide until it ceases producing an antitoxin, leading to bacterial suicide. Some bacteria can also release a large number of different, specific poisons into the environment, termed *bacteriocins*. Bacteriocins are toxins that kill their "victim" directly, instead of by inducing programmed cell death. The bacterium that produces the bacteriocin is protected from its effects by an immunity protein. Other ways for bacteria to kill each other require physical contact between the cells. A highly sophisticated system was recently discovered in which bacteria use the type VI secretion system (see below) to fight each other in deadly duels reminiscent of fencing.

Bacteriocin genes are often located in proximity to the bacterium's corresponding immunity protein genes, and in some cases, close to other genes for proteins that lyse the organism's own cell walls to release the toxins. However, it is more common for bacteria to encode specific

Bacteriocins

Bacteriocins are bacterial proteins released into the environment that are toxic to other bacteria. They are, in fact, the most powerful antibacterial agents and come in many varieties. They generally have three domains: a central domain that attaches to a receptor present in the target bacteria, another domain that helps them penetrate their target, and a domain that kills the target cell. The receptor on the target bacterium is often a nutrient receptor.

Bacteriocins from Gram-negative bacteria have a fairly narrow spectrum of activity, acting mostly against bacteria similar to themselves. They are generally able to perforate the target bacteria's membranes; some are nucleases capable of degrading DNA and RNA in the target bacteria. Bacteriocins from Gram-positive bacteria diffuse within their targets' cell walls and therefore have a wider spectrum of activity than those of Gram-negative bacteria. Gram-positive bacteria, particularly lactic acid bacteria, produce a variety of bacteriocins that are categorized into four classes: lantibiotics, small heat-stable peptides, heat-labile proteins that kill bacteria by lytic or nonlytic mechanisms, and cyclic peptides.

Not all bacteria have immunity proteins against their own bacteriocins. One exception is the bacterial RNase called *barnase* produced by *Bacillus amyloliquefaciens*, which simultaneously produces a protein called *barstar* that protects the bacillus from its effects.

transporters that allow bacteriocins to be released rather than this leakage mechanism.

The first bacteriocin was identified in *E. coli* in 1925 and named *colicin*. The second, produced by *Lactococcus lactis*, was discovered in 1927 and named *nisin*. Nisin is used as a food additive (E234) and as a preservative agent in meat and other foods. It is a very effective agent against *Listeria monocytogenes* bacteria, among many others.

Contact-dependent inhibition of growth

In addition to bacteriocins, which bacteria release into the environment to kill other bacteria that are not necessarily close by, bacteria also compete for survival with a phenomenon discovered a decade ago, the CDI system (bacterial *c*ontact-*d*ependent *i*nhibition of growth). Some bacteria carry the toxin CdiA on their surfaces, which can interact with BamA receptors on target bacteria. Upon contact between the cells, the bacteriocin cleaves to produce a toxin termed CdiA-CT that is released and penetrates the target bacterium, where it can degrade DNA and RNA or interact with a compound in the interior of the bacterium that activates the toxin. Bacteria that produce CdiA also have a protein, CdiI, that can inhibit CdiA-CT. CDI systems are not unique. Other systems, such as the Rhs system, function similarly.

Type VI secretion: attack and counterattack

Bacteria secrete proteins into the environment or pass them on directly to neighboring bacteria or eukaryotic cells via seven known types of mechanisms (designated type I through type VII secretion systems). The type VI secretion system is one of the most recently examined. It was discovered in 2006 in *V. cholerae* and has since been identified in *Pseudomonas*, which often infects cystic fibrosis patients; in *Helicobacter*, which is responsible for stomach ulcers; and in many other bacteria that infect humans. It has also been observed in bacteria that infect plants, either pathogenic (such as *Agrobacterium*) or symbiotic (such as *Rhizobium*). Type VI secretion is involved in bactericidal interbacterial interactions and in competitive growth in biofilms containing multiple species of bacteria.

The type VI secretion system involves an organelle, a nanomachine that resembles a syringe with its base in the inner membrane of the bacterium.

Type VI

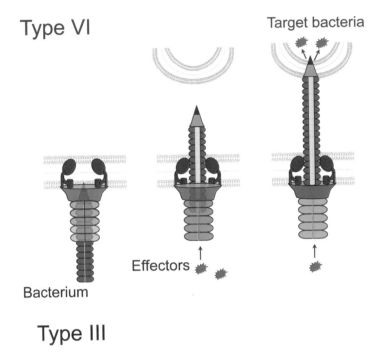

Type III

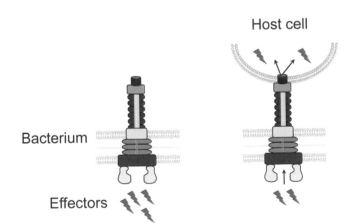

Figure 14. Type VI and type III secretion systems.

It passes through the peptidoglycan layer and its hollow piston is contractile, located either at the interior of the bacterium or extending out through the surface to penetrate into a neighboring bacterium or eukaryotic cell (Fig. 14). It injects "effector protein" toxins or enzymes that either degrade the peptidoglycan or bacterial membranes or modify actin in eukaryotic cells. It appears that type VI secretion systems can kill target bacteria just by perforating their membranes. In mixed cultures, *Pseudomonas aeruginosa* can attack and eliminate its competitor *Pseudomonas putida*.

A surprising characteristic of type VI secretion systems is that they often seem to be activated by other type VI secretion systems in what could be described as a "tit-for-tat" reflex. Bacteria under attack by a type VI secretion system will counterattack with their own type VI secretion system. This has been demonstrated in *P. aeruginosa*, which engages in an antibacterial duel with its competitors *V. cholerae* and *Acinetobacter baylyi*. *P. aeruginosa* is better able to kill other bacteria if they have a type VI secretion system too—contact with the foreign system activates the *P. aeruginosa* system. Although not enough research has been conducted to be certain what happens in polymicrobial environments such as intestinal microbiota, it is probable that type VI secretion systems are involved in homeostasis in complex microbial systems and communities.

CHAPTER 9

Human-Animal Symbioses: The Microbiotas

A revolution is shaking the field of microbiology by revealing the unexpected concept that all forms of life rely on symbiosis with bacteria—or rather with microbial assemblies. These communities are constantly changing in composition and size and play numerous important roles in the physiology and pathology of organisms; in humans, they influence our lives from the earliest stages of embryonic development until our death.

In classical microbiology, bacteria were studied in liquid or solid cultures, isolated from their natural ecosystems. However, less than 1% of the microorganisms present in common environments can be isolated by classical methods. At the end of the 20th century, environmental microbiologists began to use new molecular approaches based on high-throughput DNA sequencing to study all the bacterial populations present in an ecosystem at once. These "metagenomic" approaches allow the study of communities of bacteria and archaea present in a variety of environments, including within large animals and humans. By now, scientists have used

DNA sequencing to characterize microbial communities in humans of different origins, eating habits, and health status, before and after treatment with antibiotics, and over various lengths of time. A whole new microbiology field is emerging, revealing that microbial symbiosis exists everywhere and is critical for the survival of the organism because it both provides a variety of essential nutrients to the tissues that harbor the symbiotic organisms and plays a major role in protecting the organism against pathogens.

The term *microbiota* is used to designate the community of microbial species present in a particular environment. In contrast, a *microbiome* refers to the environment characterized by a particular microbiota, including the microbes' ensemble of genes and other characteristics, regardless of the precise species that compose it. The microbiotas produce innumerable compounds that often benefit the organism it inhabits.

The paradigm of *Euprymna scolopes* and *Vibrio fischeri*

Euprymna scolopes is a tiny squid that lives in shallow water in the Pacific Ocean. Squids spend the day burrowed into the sandy ocean bottom, and at night they hunt small shrimps that constitute their diet. Predators can spot the squid by the shadows they cast on the ocean floor in the moonlight, so the squid have evolved an ingenious means of self-protection. On the underside of each squid's body is a light organ packed with luminescent *Vibrio fischeri* bacteria. These bacteria cast a glow on the sand below, making it appear as if the squid does not cast a shadow. At the end of the night, before the squid digs into the sand for the day, it releases most of the bacteria from the light organ. Over the next 12 hours, the remaining bacteria regenerate, nourished by compounds produced within the light organ.

Researchers have recently asked whether these *V. fischeri* bacteria or the light that they produce have an impact on the squid's circadian rhythms (the internal clock that controls the sequence of functions over the 24-hour day). The laboratory of Margaret McFall-Ngai in the United States has demonstrated that the *escry1* gene in *E. scolopes*—which regulates other genes that code for the cryptochrome class of proteins that are known to be involved in circadian rhythm—is activated in the presence of *V. fischeri* but not in its absence. Similarly, if the bacteria are present but do not produce light, the *escry1* gene is not active unless the squid is illuminated at the right wavelength.

Bacterial light production is related to quorum sensing in that the bacteria emit light only when a certain cell density is reached—such as at the end of the day when the squid once again ventures out to hunt a meal.

This intriguing result shows that bacteria can control the biological rhythms of the animal that hosts them. Humans have a gene, *cry1*, that is similar to *escry1*, although it is not yet known if some bacteria regulate it. Nature is rich in innumerable symbiotic relationships neither as simple nor as visible as the *Vibrio*-squid pair. The following pages discuss numerous beneficial but nonessential symbioses in both humans and animals. It has been shown that symbiotic relationships also exist in plants. Furthermore, "endosymbiosis" has been extensively studied in insects, and symbiotic relationships may also affect other types of animals.

The intestinal microbiota

For years we have read that the human intestine harbors 10^{14} bacteria, i.e., 10 times the number of human cells that make up the body itself. An article recently called this evaluation into question but concluded that even if there are *only* as many bacteria in a person as there are human cells (10^{13}), the number is still huge! This number is the sum of the various microbiotas present in many locations of the body. The intestinal microbiota is the paradigm and one of the most thoroughly studied, but there are many other microbiotas, including those of the skin, vagina, and oropharyngeal cavity.

Our knowledge of the intestinal microbiota has advanced significantly due to metagenomic techniques that have made it possible to create precise inventories of the species and genes present. Metagenomic techniques allow for sequencing the DNA of all the organisms in a microbiota without having to isolate them. These findings significantly surpass the data obtained from studies on stool cultures, which, although performed anaerobically, only identified the subset of bacteria that were capable of growth under those conditions.

Current techniques are based on the amplification of regions of DNA that code for 16S RNA in prokaryotic ribosomes and on the complete sequencing of all DNA fragments present in a sample. These approaches have shown that the intestinal microbiota is composed mainly of five phyla of bacteria: the *Firmicutes* (60 to 80%), consisting mainly of clostridia and lactobacilli; the *Bacteroidetes* (20 to 40%); and *Actinobacteria* and *Proteobacteria* (small percentages). It also contains members of the phylum *Verrucomicrobia* such as *Akkermansia*, a bacterium capable of degrading mucus. Note that the intestinal microbiota is not in direct contact with intestinal cells, which are protected by a thick layer of mucus.

Obviously, many questions arose as soon as the first microbiotas were studied. Is the composition of a microbiota the same in all individuals? Does it change over the course of a person's life? Which factors are involved in altering its composition? And above all, what is the role of the microbiota? Numerous laboratories actively work in this new and exciting area of research, and these issues will be discussed below.

Products of the intestinal microbiota

It has long been known that bacteria throughout the digestive tract participate in digestion, and those in the intestine are involved in the final phase of the digestion. Bacteria in the intestinal microbiota produce enzymes that break down food, hydrolyze sugars, and ferment food residue into compounds that the epithelial cells in the colon can absorb. These products include organic acids such as succinate and lactate and short-chain fatty acids (SCFAs) such as acetate, propionate, and butyrate, which represent important sources of energy. SCFAs control colonization by pathogenic bacteria by regulating the expression of virulence factors that these pathogens produce; thus, they can affect the infection process at several stages. It has been shown that SCFAs impose epigenetic marks on intestinal cells, although the consequences of this are still unclear. (Epigenetic marks are modifications of the genes that do not affect the DNA sequence but may affect gene activity and expression.) Additionally, the intestinal bacteria synthesize numerous metabolites that benefit their host, such as vitamins B and K.

Evolution of the intestinal microbiota over a lifetime

The fetal intestine is sterile; it is not colonized with bacteria until birth. During the first 2 years of life, the composition of the child's intestinal microbiota is influenced by many factors, including birthplace environment, natural or cesarean birth (the vaginal microbiota influences the composition of the microbiota in the child), and above all diet (breastfeeding, milk, or other food). As the child develops, his or her microbiota will influence immune system maturation and cerebral development. The most remarkable recent advances on this subject concern the relationship that exists between the intestinal microbiota and brain development, suggesting that this microbiota might have some influence on adult behavior. Studies in mice have shown effects on anxiety and motor functions. This is why the intestine is sometimes referred to as a "second brain."

A highly diverse microbiota indicates good health, whereas a less diverse microbiota is often observed in malnutrition or illness. What does it mean for a microbiota to be diverse? Which bacteria are present? An interesting concept has recently arisen, based on a study that compared the metagenomes of 22 Europeans, 13 Japanese, and 4 Americans, a total of 39 individuals of six different nationalities. This study gave rise to the concept of "enterotypes," that is, general categories of microbiotas that are not unique to any specific country or continent. Enterotype 1 contains high levels of *Bacteroides* strains. Enterotype 2 is rich in *Prevotella* and *Desulfovibrio*. Enterotype 3 is often rich in *Ruminococcus* and *Akkermansia*, two bacterial genera that are capable of degrading mucus, the polysaccharide layer that coats the epithelial cells lining the intestine. However, although appealing at first, the enterotype concept now appears to be too rigid. A recent study, based on the analysis of a single healthy individual over the course of a year, observed that the individual's microbiota varied among several enterotypes, leading to the conclusion that the notion of distinct enterotypes is incorrect. The distribution of bacteria in microbiotas continually changes; in effect, there is a continuum of different microbiotas.

Nevertheless, aren't individual microbiotas relatively stable? Does the microbiota acquired during the first years of life characterize an individual forever? This does seem to be the case. Microbiotas are so well adapted to each individual that it appears that, even after being weakened by treatment with antibiotics, the intestinal bacteria will reestablish themselves. Bacteria are resilient creatures. Furthermore, a recent study shows that the predominant commensal bacteria are resistant to antimicrobial peptides associated with high-level inflammation and that this is due to the modification of a molecule present in cell walls of many Gram-negative bacteria, namely, their liposaccharide.

An individual's microbiota would be a new kind of identity card. Most bacteria in the intestine remain there for years. The stability of a microbiota correlates with the stability of the individual's weight. However, the microbiota will fluctuate over the course of a lifetime, affected by age and diet—which accounts for the very different microbiotas observed in different parts of the world. It appears that as people get older their microbiotal diversity decreases. Microbiotas are very different in people who are obese compared to those in lean individuals, as well as in patients with intestinal inflammatory conditions such as Crohn's disease or ulcerative colitis.

Obesity and metabolism

The microbiota has been the subject of many studies concerning obesity. Over a decade ago, Jeffrey Gordon's group at Washington University in St. Louis demonstrated that the composition of the intestinal microbiota in obese mice is not the same as in normal mice; a number of different bacterial groups vary sharply, with an increase in *Firmicutes* and a significant reduction of *Bacteroidetes* observed in the obese mice. The obese mice had an intestinal microbiome that converted food energy to body weight more efficiently. This work raises the idea that intervention aiming to change the composition of the intestinal microbiota could become a basis for new obesity treatments. Such proposed strategies have to be seen in parallel with the fact that antibiotics are still used in some countries as food additives in order to increase livestock growth. These antibiotic treatments may alter the microbiotas of the animals toward more efficient energy conversion. However, this has not yet been studied in great detail.

Several interesting studies have been conducted on pairs of twins in which one was obese and the other had a healthy weight. Fecal content from the respective research subjects was fed to mice. After ingestion of the microbiota from obese humans, each mouse's microbiota changed to become more like the microbiota they ingested, with an increase in *Firmicutes*, which led to weight gain and an increase in body fat. Since mice are coprophagous (they eat feces), a notable observation was made when obese mice and nonobese mice were reared together—the obese mice acquired *Bacteroidetes* from the microbiota in the feces of the thin mice and lost weight!

Recent work has confirmed that the acquisition of a stable intestinal microbiota occurs in mice in the early stages of life. This work also showed that treatment, even a weak dose of penicillin in early life, can introduce stable metabolic alterations leading to obesity or a predisposition to obesity and have an effect on genes involved in immunity. Epidemiological data also indicate that this is the case in humans.

A noteworthy study recently showed that food additives such as carboxymethyl cellulose or polysorbate 80 induce a mild inflammation and obesity or the metabolic syndrome associated with prediabetes and that these effects are due to changes in the composition of the microbiota.

Microbiota and the immune system

Commensal bacteria are not recognized as foreigners by the immune system. They do not induce a response from the immune system, which focuses on the destruction of invaders. The macrophages and dendritic cells of the intestine's lamina propria ignore commensal bacterial compounds—they don't "see" them—whereas they do respond to pathogenic bacteria.

However, the presence of commensals reinforces the immune response to pathogens, as if the commensal bacteria continually stimulate the immune cells to be on the alert. The microbiota triggers a mild inflammation—manifested by the production of antimicrobial peptides (type C lectins, REG3-γ and REG3-β, and alpha defensins) and the maturation of lymphoid structures—by stimulating antibody-producing cells that prevent bacteria from penetrating the mucous membrane. Most importantly, the microbiota stimulates the maturation of T lymphocytes into proinflammatory Th17 lymphocytes.

An interesting development on this last observation comes from studies of the microbiota in mice susceptible to *Citrobacter*, a pathogen similar to *Escherichia coli*. In these mice, the bacterium responsible for the generation of Th17 lymphocytes was identified as "*Candidatus* Arthromitus," a strictly anaerobic, segmented filamentous bacterium (SFB) similar to *Clostridium*. Thanks to a recent technique for culturing this previously uncultivable organism in cell cocultures, the way is now clear toward determining the role this commensal bacterium plays in the stimulation of a physiologic inflammation in the gut.

SFB in the intestine

Two independent studies have shown that filamentous bacteria in the *Clostridiaceae* family, termed segmented filamentous bacteria, or SFB, induce differentiation of Th17 cells in the lamina propria of the intestine. These Th17 cells then secrete proinflammatory interleukin-17 (IL-17) and IL-22 cytokines. When IL-22 causes the production of antimicrobial proteins, mice treated with SFB are more resistant to infection with *Citrobacter rodentium*, a bacterium in mice identical to enteropathogenic *E. coli* (EPEC) in humans. SFB were considered until recently to be uncultivatable in a lab setting, but recent developments in cultivation techniques will undoubtedly lead to significant progress in our knowledge of the modes of stimulation of the type B immune cells by SFB.

Microbiota, pathogenic bacteria, and dysbiosis

Mice that lack an intestinal microbiota—either germfree or treated with antibiotics—have been shown to be much more susceptible to infection than normal mice. There are several ways that the intestinal microbiota provides protection. Commensal bacteria occupy niches in the intestine that pathogens would otherwise invade. They compete against outsiders for the available nutrients. Furthermore, they can produce antibacterial substances such as antibiotics or bacteriocins or release active lytic bacteriophages against invaders.

In order to attach to epithelial cells, pathogenic bacteria must traverse the thick layer of mucus that covers the intestinal epithelium. The intestinal microbiota affects this as well. The host generates mucus in response to bacterial compounds that the microbiota produces. Axenic mice (lacking a microbiota) have a much thinner mucus layer than do normal mice, but the administration of specific bacterial compounds restores normal mucus production.

It seems that certain bacteria contribute more than others to strengthening the intestine's barrier function. *Bifidobacterium longum* secretes peptides that control intestinal permeability, along with SCFAs that raise the host's intestinal defenses.

While the intestinal microbiota protects intestinal tissue from pathogenic aggression in several ways, it has been observed that some bacteria in the microbiota can favor infection. It was recently shown in mice that the presence of the commensal *Bacteroides thetaiotaomicron* leads to the production of succinate, which can have a major effect on infection. In intestinal infection with *C. rodentium* in mice, for example, the infection increases in the presence of *Bacteroides*. This influence of the microbiota on local infection may explain why some people frequently suffer from enteritis while others never do.

Another study has shown that *Clostridium difficile* utilizes the succinate produced by *B. thetaiotaomicron*. *C. difficile* transforms this succinate into butyrate by fermenting carbohydrates such as sorbitol, thereby increasing its growth and favoring colonization.

Dysbiosis and fecal transplants

More and more attention is focusing on dysbiosis—microbial imbalance in a microbiota, i.e., a change in the distribution of different types of bacteria—which may allow the growth of both pathogenic bacteria and bacteria that are normally not pathogenic but benefit from lowered competition for nutrients usually consumed by commensals. Dysbiosis favors the persistence of infections and thus their transmission from host to host.

Given that antibiotic resistance is increasing and few new antibiotics are being developed, the value of a balanced microbiota in the course and possible transmission of infections indicates that treatments aimed at the reconstitution of a healthy microbiota can offer strong promise. Such strategies have already been tested in patients with recurrent *C. difficile* infection. When given microbiota transplants from healthy donors, the patients' success rate was higher than for treatment with vancomycin alone. That being said, there is still a risk of unsuitable microbiota being transplanted as well. Which species present in a normal microbiota best prevent the growth of bacteria such as *C. difficile*? To respond to this question, mice were treated with different antibiotics and then exposed to *C. difficile*. Those that resisted the infection featured *Clostridium scindens* in their microbiota, a species that is capable of degrading bile and producing metabolites that counteract *C. difficile* infection. This led to the conclusion that enriching microbiota with *C. scindens* or secondary bile acids could increase the effectiveness of fecal transplants.

Microbiota and the human diet

Recent experiments in humans have shown that the intestinal microbiota can respond rapidly to changes in diet as part of a lifestyle change. Two different diets were tested in adult participants over a 5-day period: an animal-based diet composed of meat, eggs, and cheese and a vegetarian diet rich in grains, fruits, and vegetables. The meat-based diet increased the level of bile-tolerant bacteria (*Alistipes, Bilophila,* and *Bacteroides*) and decreased the number of *Firmicutes* that metabolize plant polysaccharides (*Roseburia, Eubacterium rectale,* and *Ruminococcus bromii*). The differences in microbial activity were similar to those observed between herbivorous and carnivorous animals. Another study revealed diet-linked changes in microbiota as partially responsible for the decline of health during aging.

Microbiota and circadian rhythms

The circadian rhythm is known to play an important role in regulating digestive system functions, particularly nutrient absorption, cellular proliferation, motility, and metabolic activity. This is why, for example, night-shift workers may experience intestinal problems. We know that the microbiota is responsible for maintaining intestinal equilibrium. Recent studies have shown that the circadian clock controls intestinal homeostasis regulation by the microbiota, because Toll-like receptors responsible for the physiological inflammation discussed above ("Microbiota and the Immune System") are regulated by the circadian rhythm. Problems in the circadian rhythm or intestinal dysbiosis caused by antibiotic treatments or infections, for example, are both associated with obesity, metabolic syndrome, and chronic inflammatory illnesses. Several recent studies have shown that upsetting the circadian rhythm and eating meals irregularly have a significant impact on the composition of the intestinal microbiota, which can result in metabolic disorders or inflammatory diseases.

These data suggest that previous studies on the composition and role of the microbiota should be reevaluated in light of these findings. The time of day when samples are collected and how recently meals were consumed by participants are important factors to take into consideration when conducting research on the intestinal microbiota.

Skin microbiota

The skin is the body's largest organ, with an average surface area of 1.8 square meters. It is a physical barrier that prevents foreign pathogens from entering the body. There are roughly half a million bacteria per square centimeter, totaling roughly 10^{10} (10 billion) bacteria living on the surface of our skin. Our skin's surface varies greatly in pH, temperature, hydration, levels of sebum, and topography. The skin is less hospitable and much less rich in nutrients than the intestine. Nonetheless, it is equipped with a highly sophisticated immune surveillance system that is the result of the combined activities of epithelial cells, lymphocytes, and antigen-presenting cells in the dermis and epidermis. In contrast to the intestine, the skin microbiota doesn't seem to play a role in the establishment of the immune system. On the other hand, it does control the expression of many factors of innate immunity—that is, the immediate and nonspecific response to a given pathogen—such as antimicrobial skin peptides, chiefly cathelicidins and beta-defensins.

The skin microbiota also augments other defense mechanisms, such as the complement system, which is involved in pathogen opsonization (a process by which the target surface is marked by antibodies, complement proteins, or other molecules for ingestion and destruction by a phagocyte) and inflammatory responses that help eliminate pathogens. The microbiota also increases levels of IL-1, a cytokine (small protein) involved in immune system responses. Like the intestinal microbiota, the skin microbiota is associated with a variety of inflammatory disorders. These include chronic conditions such as atopic dermatitis (eczema), psoriasis, and acne. The prevalence of eczema has more than doubled in industrialized countries. Eczema is often associated with *Staphylococcus aureus*, but the role of the skin microbiota in this condition is not yet well documented. Psoriasis is attributed to *Streptococcus* species, but more studies are needed on the influence of microbiota on this condition and on acne as well. Acne has been attributed to *Propionibacterium acnes*, but these bacteria are equally present on the skin of people who have acne and those who do not! Perhaps these bacteria can become commensal or pathogenic depending on their host. Data should be revisited in light of the current understanding and discoveries on the role of the skin microbiota.

Vaginal microbiota

The vaginal microbiota, sterile at birth, eventually develops into a community composed mainly of four lactobacilli. The vaginal environment is thus acidic due to the lactic acid produced by these lactobacilli. The composition of this microbiota becomes more diverse during puberty. Many studies are being undertaken to determine what role, if any, the vaginal microbiota plays in fertility.

Intestinal microbiota of termites

Termites are insects that digest the cellulose and lignocellulose found in wood, making their intestinal microbiota particularly interesting. Termites make up 95% of the "insects" biomass present in tropical soil and in the African savanna. Although there are over 3,000 species of termites, only a few pose a threat to structures made from wood, but these few nevertheless have a huge impact on tropical agriculture and economics. Termites that break down wood occupy a central place in the carbon cycle because of their intestinal microbiota, consisting of bacteria, archaea (a hundred times less numerous than bacteria), and flagellate protists. These microorganisms degrade plant fiber and ferment the derivatives into acetate and methane, with hydrogen as an intermediate that can be stored by the flagellated protists in organelles known as *hydrogenosomes*. The methane is produced by methanogenic archaea, such as *Methanobrevibacter*.

Cellulose is broken down in a synergistic interaction between the termites and their own microbiota. The intestinal microbiota synthesizes numerous compounds that provide nutrients for the termites. The mineralization of humus compounds in the intestines of certain termite species that get their nutrients from soil contributes to the nitrogen cycle. The amazing efficiency of termite intestines as bioreactors is a promising model for industrial conversion of lignocellulose into microbial products and for the production of biofuels. That said, many of the precise details and mechanisms involved are not known and may be difficult to define. Techniques used in metagenomics (the study of genetic material in environmental samples without the need to isolate individual bacteria) and metatranscriptomics (the similar study of RNA transcripts) will facilitate further research. However, many microorganisms in both human and termite microbiota are not cultivatable *in vitro*, which is a major obstacle for the comprehensive understanding of phenomena.

Composition of microbiotas: signaling molecules and quorum sensing

As previously discussed, bacteria produce signaling molecules that they use as a language of communication between bacteria and by which they recognize each other and assess bacterial concentration. In antibiotic-treated mice, the *Firmicutes* proportion decreases, but if a colibacillus, which produces the molecule AI-2 that all species of bacteria can recognize, is introduced into the intestine, the *Firmicutes* reestablish their numbers. This shows that during antibiotic treatment, quorum-sensing molecules can be used to restore a microbiota rich in "beneficial" *Firmicutes*.

Longevity and microbiota

Studies conducted on the microbiota of the fruit fly *Drosophila melanogaster* are in the process of shedding light on the role of the microbiota in longevity. It is quite possible that someday we will be able to lengthen human life expectancy by controlling the composition of the human intestinal microbiota. Although the connection between the *Drosophila* findings and those in humans is still hypothetical, many studies in humans indicate that good health is linked to having a diverse microbiota and that a diverse diet helps maintain a healthy intestinal microbiota.

CHAPTER 10

Bacterium-Plant Symbioses: Microbiotas of Plants

Like humans, plants play host to bacterial communities of various compositions depending on their location on the plant. These communities are generally restricted to just a few bacterial phyla: *Actinobacteria, Bacteroidetes, Firmicutes,* and *Proteobacteria.* The root microbiota acquires its members from the surrounding soil but also depends on factors from the plant itself. In the same way, the microbial communities present on the surface of plant leaves depend on the substrates available. Like those in leaves, root microbiotas provide protection from pathogens, but they also perform other roles, including the acquisition of beneficial nutrients from the soil to sustain the plant. Plant microbiotas appear to be involved in a form of mutualism important to the plant's growth and well-being. They allow the plant to adapt to a wide variety of environments.

Research over the past 30 years has focused on binary interactions between plants and bacteria and has mainly been concerned with either plant diseases, particularly bacterial virulence factors and molecular mechanisms that stimulate the host's immune responses, or the symbiotic relationships between leguminous plants and nitrogen-fixing bacteria involved in the formation of root nodules. In both cases, disease symptoms or macroscopically visible structures have been observed as the result of parasitic or mutualistic interactions that can be considered extremes along a continuum of interactions between plants and microorganisms and are now being intensively studied.

Recent studies have shown that healthy plants harbor an impressive array of microbes. Just as humans can be said to be "superorganisms"—living beings characterized by the variety of microbiomes that inhabit the human body—so the assortment of microbiota within plants could define them as "superplants." Expanding research is focusing primarily on roots and leaves.

Microbes and roots: the underground

Soil is one of the richest ecosystems on the planet. Its diversity is made up of several phyla including *Acidobacteria, Actinobacteria, Bacteroidetes, Chloroflexi, Firmicutes,* and *Proteobacteria.* Soil bacteria are present in the *rhizosphere,* the environment surrounding roots, and in the *endosphere,* the environment inside of roots. The rhizosphere and the endosphere have different compositions, showing that edaphic factors (abiotic factors coming from the soil) and factors from the plant regulate bacterial microbiotas present within or in close proximity to roots.

Plant cells on the surface of roots make up the *rhizodermis.* In a process called *rhizodeposition,* these cells release a variety of compounds, including organic acids, inorganic ions, phytosiderophores (compounds that sequester iron in plants), sugars, vitamins, amino acids, purines, nucleosides, and polysaccharides, that are involved in bacterial deposits on and around roots. Some of these cells can desquamate (peel away), but remain alive after separation, and can attract bacteria. Some of the bacteria attracted by roots provide them with favorable growth conditions, it is the case for *Pseudomonas fluorescens* in tomatoes; however, pathogens can also enter into a plant through its roots. *Ralstonia solanacearum,* for example, is attracted by various amino acids, organic acids, and exudates released by the tomato plants it targets. Some mutants of this bacterium were shown to have lost either chemotaxis or the ability to perceive exudates, and were thus avirulent. However, these studies were not conducted in the presence of a natural microbiota and therefore need to be refined as technological advances make it possible to examine specific bacteria in the presence of soil microbiotas from their natural environments.

Microorganisms in the soil can either help or hinder plant diseases that originate in roots. Rhizosphere microbiotas compete with pathogens with varying degrees of success depending on the composition of the soil. Certain soils are therefore more favorable for certain plant diseases. However, it is ultimately a plant's roots that control the rhizosphere's composition and consequently the plant's susceptibility to diseases.

For at least 10,000 years, humans have cultivated plants as food for needs and gustative preferences. This has undoubtedly altered the microbiota of domesticated plants compared to wild plants, which are frequently exposed to much more challenging environments.

Nitrogen fixation: an example of facultative symbiosis

Symbiosis refers to cohabitation that is beneficial for each party involved. It can be either facultative or obligate. Most frequently, symbiosis is *facultative*, in which a partner does not require the other to survive. However, *obligate* symbiotic relationships have developed over the course of evolution and are particularly common in insects.

In plants, obligate symbiosis is rare. *Rubiaceae*, a certain family of flowering plants that includes coffee and jasmine, is an exception. These plants are associated with certain strains of *Burkholderia* that have fewer genes and a smaller genome than other *Burkholderia* species. *Rubiaceae* plants deprived of these bacteria will not attain maturity.

The most beneficial plant-bacteria cohabitation known is between legumes and *Rhizobium leguminosarum*, in which the bacteria, occupying small nodules on the roots, transform nitrogen from the air and soil into ammonia. The trapped nitrogen enhances the plant's growth while the bacteria obtain carbon and energy from the plant. As most soils are poor in nitrogen, requiring the extensive use of fertilizers, bacterial nitrogen fixation can be of significant agronomic and economic importance. *Rhizobium*-leguminous interactions bring into play mutual recognition of signals diffused between the plant and its symbiont.

Bacterial communities and the phyllosphere

The *phyllosphere* comprises all the parts of the plant above the ground, including stems, leaves, flowers, and fruits. Leaves make up the majority of the

Nitrogen fixation and root nodule formation

Certain bacteria such as *Sinorhizobium meliloti* establish symbiosis with a plant to efficiently fix the nitrogen necessary for plant growth. In rich soils, where the plant has adequate nutrition on its own, the plant and bacteria live independently. However, if nitrogen drops to a limiting level, the stressed plant produces flavonoids. These compounds stimulate the neighboring *Rhizobium* bacteria to express genes and nodulation, or Nod, factors that promote what is called the *infection* of the plant roots by the bacterium (this term is somewhat misleading, because it is not an infection in the sense of causing an illness, but much the opposite!) and development of the nodule. In some cases, bacteria can enter through injury to the plant. More often, they attach to the tips of root hairs that then curve inward to form a sort of pocket. Called an *infection tube*, this structure penetrates the cortex and allows the bacterium to enter the plant via invagination of the plant's cell membranes. At this point the cell cycle speeds up, cells proliferate and differentiate, and nodules begin to form around the bacteria, creating structures called *symbiosomes*.

As the nodule develops, the physiology of the bacteria adapts to the intracellular milieu of the host cells and to endosymbiotic existence. The structure of the nodule offers semianaerobic conditions ideal for nitrogen fixation. The low levels of oxygen within the symbiosome activate bacterial genes involved in nitrogen fixation and an altered cellular metabolism. The morphology and physiology of the bacteria can vary significantly depending on the host. For some legumes, the bacteroid state is reversible; in others, the differentiation process is irreversible.

Nitrogen fixation is accomplished by an oxygen-sensitive enzyme, nitrogenase, a tetramer that contains molybdenum and iron. Homocitrate, a component of the molybdenum-iron cofactor, is also required. Although the bacteria produce nitrogenase, they cannot synthesize homocitrate (rhizobia lack homocitrate synthase) and must acquire it from the host. This complementarity illustrates the essential partnership between plant and bacteria for nitrogen fixation.

phyllosphere. The microbiota here is dominated by bacteria but also includes microorganisms such as archaea and fungi. They are subjected to frequent changes in temperature, humidity, and ultraviolet radiation and lack access to many nutrients. They participate in the carbon and nitrogen cycles and protect the plant from pathogens.

As with the intestinal microbiota or the rhizosphere, the bacterial phyla of the phyllosphere are limited to *Actinobacteria*, *Bacteroidetes*, and *Firmicutes* as well as *Proteobacteria*, which are dominant. Although the air and aerosols are bacterial sources, the concentration of bacteria in the air is less than in the ground. Too few studies have been conducted to be able to determine the factors that influence the selection of bacteria in the phyllosphere.

Bacteria and plant growth

In addition to their ability to fix atmospheric nitrogen in nodules, bacteria are able to make phosphate soluble, allowing plants to absorb it. Bacteria can also produce phytohormones, such as auxins, that are involved in different aspects of plant growth and development. Direct contact between plants and bacteria is not always necessary; some microorganisms release volatile organic compounds. *Bacillus subtilis*, which produces the growth-promoting compounds 3-hydroxy-2-butanone and 2,3-butanediol, is a well-documented example.

Like the intestines of humans and animals, the root system is critical to the growth of the plant. Root microbiotas regulate the absorption of nutrients and protect the plant from pathogen invasions. Although it may take years to adequately study the interactions of different microbiota, soil composition, and plant behavior, the data collected will be critical to maintaining the environmental equilibrium and improving 21st century agriculture, thus improving food quality.

CHAPTER 11

Endosymbiotic Relationships

Symbiotic relationships between bacteria and eukaryotes (humans, animals, and plants) are generally beneficial to at least one of the partners involved. *Facultative* symbioses can arise, fade, and sometimes recur. *Obligate* symbioses, however, become established by evolution over time. They are particularly common in insects: 10 to 12% of insects carry intracellular symbionts, known as *endosymbionts*. Endosymbiosis has contributed to the evolutionary and ecological success of many insect species, providing properties that allow them to adapt to niches that would be otherwise inaccessible. The first endosymbionts are thought to have been mitochondria and chloroplasts in eukaryotic cells. In this process, a cell with a nucleus would have established an obligate symbiotic relationship with a photosynthetic bacterium, ultimately evolving into plant cell chloroplasts, or with a nonphotosynthetic bacterium, ultimately becoming the mitochondria found in all eukaryotic cells.

A close-knit couple: the pea aphid and *Buchnera* bacterium

The most thoroughly studied obligate symbiotic relationship is between *Acyrthosiphon pisum*, the pea aphid, and *Buchnera aphidicola*. It is an obligate relationship for both the aphid and the bacterium, as the bacterium is only found in this aphid. Study of the genomes of the pea where these aphids live and of the bacteria have shown this symbiosis to be mainly nutritional.

B. aphidicola is a member of the family *Enterobacteriaceae*, well known for *Escherichia coli*. Genomic studies have shown that over the course of evolution, *Buchnera* has lost many genes—a characteristic of many obligate intracellular bacteria. These genes once allowed it to live independently in the environment but are unnecessary for its intracellular lifestyle. Examples include genes for the synthesis of compounds that the bacterium can obtain from its host, others for the liposaccharides that free-living enterobacteria require for cell structure, and several genes used for anaerobic respiration and synthesis of amino sugars and fatty acids. The result is one of the smallest known genomes for a living organism (652 kilobases) and one of the most stable.

Aphids are the cause of considerable agricultural losses. It is estimated that aphids infest one-fourth of all plant species in temperate climates, including nearly every plant cultivated for food. A classic historical example is the *Daktulosphaira vitifoliae* aphid, more commonly known as *Phylloxera vastatrix*, which decimated French vineyards in the 19th century. (These vines were later "reconstituted" by means of grafting onto imported American rootstock that was resistant to the aphids.) Aphids feed on plants by puncturing them with their syringe-like mouths and sucking out the sap. This deprives the plant of a large portion of its sap and causes damage that can make it easier for harmful viruses and bacteria to enter the plant. The sap provides the aphids with sugar, but it contains almost none of the amino acids the aphids require. *B. aphidicola* provides some of these necessary amino acids. Studies have shown that aphids raised on a diet lacking these amino acids can grow and reproduce, but if they are treated with antibiotics to kill their bacterial allies, the aphids cease growing and die. This demonstrates how critical the bacteria are to the aphids' continued survival.

In the aphid, the *Buchnera* bacteria are located inside large cells called *bacteriocytes*. Each bacterium in a bacteriocyte is surrounded by a membrane from the host, forming a vacuole called a *symbiosome*. It is thought that the amino acids produced by the bacteria are released from this vacuole and then absorbed by host cells. Each adult aphid contains several million *Buchnera* organisms.

The *Buchnera* symbionts are able to biosynthesize some, although not all, of the amino acids the aphid requires. The host aphid provides the bacteria with energy, carbon, and nitrogen. Glutamine and asparagine are amino acids that are abundant in the phloem, the plant conductive tissue for sap. When the aphid ingests them, they are transported to the bacteriocytes, where the bacteria transform the asparagine into aspartate. The aspartate in turn is transaminated to oxaloacetic acid by aspartate transaminase, releasing glutamic acid. Glutamine too is converted to glutamic acid. The bacteria utilize the nitrogen in glutamic acid to produce other amino acids that are used by the aphid. Thus, these amino acids are produced by a metabolic collaboration between the aphid and the bacteria, a conclusion confirmed by genomic data.

The aphid's genome has been the focus of much research. The aphids lack genes involved in immunity, which highlights why the acquisition of *Buchnera* by the aphids has been so successful. Although gene transfer does not occur between *Buchnera* bacteria and aphids, it does occur between the aphids and other bacteria. These genes are clearly expressed in the aphids' bacteriocytes, suggesting that they play an important role in symbiosis; however, practical studies of these genes are still in the early stages.

Aphids also sometimes have secondary endosymbionts in their intestines, in tissues surrounding the bacteriocytes or in other bacteriocytes. Pea aphids have three of these bacterial symbionts, *Hamiltonella defensa, Regiella insecticola,* and *Serratia symbiotica.* It is possible that these bacteria are peripherally involved in the mutualism between the pea aphid and the *Buchnera* bacteria: in short, that they affect the couple's relationship!

Other insects and other symbioses

Endosymbiosis is also common in the tsetse fly, which takes its blood meals from mammals. This fly has its own endosymbiont, *Wigglesworthia glossinidia brevipalpis,* that, like the *Buchnera* species in aphids, is found in highly specific cells and provides the insect with vital compounds including vitamins and other cofactors.

Carpenter ants, *Camponotus fellah,* carry the endosymbiotic bacterium "*Candidatus* Blochmannia"; however, antibiotic treatment does not appear to affect the ant's survival. The role of this endosymbiont is not yet known. The communal life of ants is so specific and complex that it will take time to elucidate all of its intricacies.

When bacteria control fertility

Buchnera bacteria are transmitted from one generation to the next via the aphid's oocytes. This is also the case for the *Wolbachia* species, an endosymbiont with astonishing properties that is even more widespread than *Buchnera*.

Wolbachia bacteria are present in 60% of all species of insects, including many mosquito species, particularly those who prey on humans—although not in *Aedes aegypti*, the main mosquito implicated in transmission of dengue virus. *Wolbachia* is also present in 47% of the *Onchocercidae* family of nematodes or filarial worms.

Studies in *Wolbachia* have focused primarily on the phenomenon of *cytoplasmic incompatibility* and its possible use for combating mosquito-borne illnesses (Fig. 15). When a male mosquito infected with *Wolbachia* mates with an uninfected female, the offspring are not viable. Furthermore, because the bacteria are transmitted by the female's ovaries, if the female is infected her offspring are also infected. These two factors—cytoplasmic incompatibility and transmission by females—increase the transmission rate of *Wolbachia* in offspring and result in generations of mosquitoes infected with *Wolbachia*. It has been observed that some *Wolbachia* species can completely eradicate males in certain insects: while infected female embryos develop normally, infected male embryos are not viable. This finding has been investigated as a way of reducing mosquito populations.

Another surprising fact is that *Wolbachia* can feminize woodlice! The presence of *Wolbachia* in fertilized woodlouse eggs induces *genetic differentiation* of male eggs into functional, infected females that produce infected offspring.

Infection with *Wolbachia* affords certain mosquitoes resistance against dengue virus, Chikungunya virus, yellow fever virus, West Nile virus, and even the parasites *Plasmodium falciparum* or *Plasmodium vivax*. Consequently, the idea of releasing *Wolbachia*-infected mosquitoes in the environment to prevent spread of viral or parasitic diseases has emerged and has been used experimentally (see below).

Bacteria and worms

Wolbachia bacteria have also been detected in roundworms responsible for illness in humans and domesticated animals. *Brugia malayi*, a filarial nematode transmitted by mosquitoes, is responsible for lymphatic filariasis or elephantiasis, a gruesome if not usually fatal disease characterized by edemas in the extremities. Dog filariasis is transmitted by ticks. It was thought until recently that 89.5% of filarial roundworms carried *Wolbachia*, but this

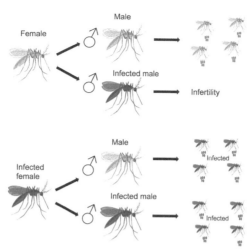

Figure 15. Cytoplasmic incompatibility. Mating between an uninfected female mosquito and an uninfected male produces offspring; however, if the male is infected with a member of the *Wolbachia* species, no offspring are produced. When an infected female mates with any male (infected or uninfected), offspring are produced and all are infected because *Wolbachia* is transmitted by female cells.

figure has now been shown to be much lower, just 37%. It is nevertheless interesting to note that although certain filarial worms do not carry *Wolbachia*, they seem to have acquired *Wolbachia* genes. This hints at a symbiotic relationship between *Wolbachia* and the worm, which has evolved into gene capture.

In filarial parasites, as in insects, *Wolbachia* passes from generation to generation through the female germ cells called *oocytes*. The relationship between the roundworm and *Wolbachia* seems to be mutualistic, with both parties benefiting from the symbiosis. This has been demonstrated by treating filariasis with antibiotics to kill the *Wolbachia*; the worms then die as well. Sequencing of the *B. malayi* genome has revealed that *Wolbachia* has genes necessary for synthesizing heme and riboflavin, which are lacking in the parasite. In addition to enabling the life and reproduction of *Brugia*, *Wolbachia* plays a role in filarial disease by causing inflammation and immune suppression in the host, as well as a size increase in lymphatic vessels in the case of lymphatic filariasis.

Ivermectin and albendazole are commonly used to treat filariasis. Ivermectin is a derivative of avermectin (which was discovered by the 2015 Nobel laureates William Campbell and Satoshi Omura), and albendazole (invented by Robert J. Gyurik and Vassilios J. Theodorides) is frequently used in conjunction with it. However, these drugs can require several years of treatment and have led to resistance. Thus, for the past dozen years, the recommended course of treatment for filariasis has been antibiotic therapy to destroy the *Wolbachia*, even though it still requires 4 to 6 months and is not recommended for children younger than 9 or for pregnant or nursing women.

Bacteria in cell nuclei and mitochondria

Intracellular bacteria, including the endosymbionts discussed above as well as many pathogenic bacteria, often take up residence in the cytoplasm of the cells they occupy, either free or enclosed in a vacuole.

That said, it appears that some bacteria have succeeded in creating mechanisms that allow them to live in the most critical part of a eukaryotic cell, its nucleus. There they are sheltered from the cell's cytoplasmic defenses and are close to the cell's DNA, which they can potentially manipulate. Intranuclear bacteria are generally associated with unicellular eukaryotes, such as paramecia or amoebae. One of the most studied intranuclear bacteria is *Holospora*, which infects paramecia and is similar to *Rickettsia*. Of the two nuclei of the paramecium, some *Holospora* species infect the macronucleus and others infect the micronucleus. *Holospora* is also found in arthropods, in marine invertebrates, and even in mammals.

Mitochondria are organelles present in nearly all eukaryotic cells except for red blood cells, which also have no nucleus. They generate ATP. It is now widely accepted that mitochondria have evolved from intracellular bacteria! They contain their own DNA, which includes a restricted number of genes that closely resemble bacterial genes, particularly those in *Rickettsia*. A surprising fact is that a bacterium—just a single species so far—has been discovered inside mitochondria. Indeed, *"Candidatus* Midichloria mitochondrii" is an endosymbiont of mitochondria in the tick *Ixodes ricinus*, which can carry the *Borrelia burgdorferi* bacteria responsible for Lyme disease. We will surely discover more bacteria that make their homes within mitochondria.

PART III

The Biology of
Infections

CHAPTER 12

Pathogenic Bacteria, Major Scourges, and New Diseases

Infectious diseases occur when a pathogenic agent (bacterium, virus, parasite, or fungus) succeeds in multiplying in an organism. Generally the agent is not present before the beginning of the illness, although members of the normal bacterial microbiota can cause *opportunistic* infections if the host's status suddenly changes, that is, if its natural immunity is impaired by immunosuppressive therapy, another infection, injury, or excessive fatigue or stress. However, most diseases are of human or, more often, animal origin. Zoonotic diseases, which are diseases transmitted from animals to humans, are most frequently transmitted by insects (such as fleas, flies, mosquitoes) or arthropods (such as ticks) that either carry the pathogen themselves or are vectors for infections in other animals (rats, various rodents, and domestic animals).

Bacterial diseases can be caused by bacteria from all genera, Gram negative or Gram positive, spore-formers or not, intracellular or not. In a single taxonomic bacterial family, some genera and species may be pathogenic while others are harmless or even beneficial. Pathogenic bacteria may be described as those that produce factors that allow them to resist their host's immune defenses and multiply in the host—whether in the host's fluids such as blood and cerebrospinal fluid, on mucosal surfaces such as the intestine, or in specific organs like the nasal cavity and the lungs—and thereby harm the well-being of the host.

The great scourges of humanity

Plague and other yersinioses

The bubonic plague is caused by *Yersinia pestis*, a bacterium first identified by Alexandre Yersin during an outbreak in Hong Kong at the end of the 19th century. Although the plague has been known since antiquity, the first recorded epidemic, referred to as the Justinian Plague, occurred in the 6th century in the Byzantine Empire. Since then, this disease has caused uncountable millions of deaths. The plague is transmitted by infected fleas carried by rats; the fleas jump to humans and transmit the bacterium in their bite. Symptoms appear a week after the bite and may include adenopathy and buboes, terms that describe swollen lymph nodes in the inguinal (upper thigh) as well as in the neck, armpits, and upper femoral region. Left untreated, the infection can quickly result in dehydration, neurological damage, and death. The pneumonic plague, which attacks the lungs, and septicemic plague, infecting the blood, are less common than the bubonic plague and can sometimes be transmitted from human to human. These versions can also prove fatal within days.

In 1347, the "Black Plague" swept Europe, killing 25 to 50% of the population. Plague outbreaks continued regularly in Europe (the Great Plague of London, the Plague of Marseille) until the bacillus responsible was finally identified, allowing the development of a vaccine. There are currently several thousand documented cases of the plague annually, 90% of which are in Africa, most notably in Madagascar and the Democratic Republic of Congo. In the United States, there are about a dozen cases per year, generally in the western United States. The World Health Organization (WHO) has noted recently that the illness is on the increase.

Rigorous studies analyzing the genomes of *Y. pestis* strains from around the globe have shown that the bacterium originated in Asia. The common ancestor of all the strains examined was identified in China, dating from 2,600 years ago.

Yop proteins and type III secretion systems

In addition to *Y. pestis*, two other *Yersinia* species are well known: *Y. enterocolitica* and *Y. pseudotuberculosis*, both enteropathogenic bacteria. *Y. enterocolitica* causes gastroenteritis, particularly in children. *Y. pseudotuberculosis* is responsible for mesenteric adenitis in humans, but it primarily affects animals.

In the late 1980s, these bacteria played a key role in the discovery of the type III secretion system (see Fig. 14). Researchers noted the major role played by Yop proteins in *Yersinia* infections. *Yersinia* bacteria release these proteins into the environment or inject them directly into mammalian cells. Studies of this mechanism revealed the existence of a sophisticated protein nanomachine that was named the type III secretion system or TSS3. This system is also present in other bacteria such as *Salmonella, Shigella*, and *Escherichia coli*.

The plague is notorious as one of the most dreaded diseases in history. Patients with plague were generally quarantined or isolated. Because no one knew what caused the plague or how it was transmitted, a variety of strategies to prevent it were tried, each as ineffective as the next. There was even a 27-kilometer-long wall built in the Vaucluse Mountains to protect the local villages from the plague of Marseille (1720). Doctors tried to protect themselves from infection by wearing masks with long birdlike beaks that contained medicinal herbs. Contemporary paintings from Venice show these masks were white.

Aminoglycoside antibiotics (streptomycin and gentamicin) are still effective against the plague, although, as for other bacteria, antibiotic-resistant *Y. pestis* strains have appeared.

Leprosy and *Mycobacterium leprae*

Like the plague, leprosy is an illness known since antiquity, mainly in China, Egypt, and India. Like those infected with the plague, lepers were feared and often rejected and excluded, even though leprosy is not very contagious. Leprosy develops slowly because the bacteria multiply extremely slowly (the bacteria's doubling time is 10 to 15 days in armadillos, one of the rare small mammals in which the bacteria can be cultivated). It spreads from human to human and, as has recently been shown in the United States, can also be passed from armadillo to human. The bacterium responsible for leprosy, *Mycobacterium leprae*, was discovered by the Norwegian Gerhard Hansen in 1873; hence it is also known as Hansen's disease. It primarily affects the peripheral nerves, skin, and mucous membranes. The bacterium multiplies in nerve cells called *Schwann cells*, which results in their destruction and the loss of feeling in extremities; in the skin, the bacterium causes granulomas and tissue destruction. Once incurable, leprosy now can be treated with a multidrug regimen of dapsone, rifampin, and clofazimine over the course of 6 months to a year. Over the past 20 years, more than 14 million patients have been cured of leprosy. There are still around 200,000 leprosy patients in the world, in Brazil, India, Indonesia, Madagascar, and several other countries, with roughly 200,000 new cases of leprosy annually in the world. The WHO has launched a strategy for the complete eradication of leprosy that seems promising.

Tuberculosis and *Mycobacterium tuberculosis*

Tuberculosis, also one of the oldest known diseases, is the second most deadly infectious disease after AIDS (caused by the human immunodeficiency virus, or HIV). In 2015, as reported by the CDC, 10.4 million people developed tuberculosis, and 1.8 million died from it. This, however, represents a significant drop of 45% in the mortality rate since 1990. A third of the world's population carries the bacterium, although only 10% develop the illness. It affects primarily individuals with weakened immune systems.

Tuberculosis is caused by *Mycobacterium tuberculosis*, the bacillus that was discovered by Robert Koch in 1882 and is still sometimes referred to as *Koch's bacillus* or *KB*. It spreads from human to human via airborne droplets from infected individuals. Although pulmonary tuberculosis, once called *phthisis*, is the most widespread form of tuberculosis, *M. tuberculosis* can also affect other parts of the body, including the bones, kidneys, intestines, genitals, meninges, adrenal glands, and skin. Before antibiotics became widespread in the 1950s, tuberculosis was treated with sunshine and fresh air in sanatoria or by surgery. It is now treated with four antimicrobial agents over the course of 6 months or longer. The emergence of multidrug-resistant (MDR) and extremely drug-resistant (XDR) strains has allowed tuberculosis to make a very disturbing comeback. Although it is now less common in developed areas, it is still a major problem in developing countries, and it is of particular concern worldwide in immunosuppressed individuals such as those with HIV/AIDS.

A vaccine for tuberculosis has existed since 1921. This vaccine—called *bacillus Calmette-Guérin* or *BCG*, named for the two scientists at the Pasteur Institute of Lille who developed it—is a "live attenuated" strain of *Mycobacterium*

Mycobacterial genomics

Like *M. leprae, M. tuberculosis* replicates extremely slowly, with doubling times of 2 weeks and 20 hours, respectively. This has hampered research on both of these diseases. In 1998 and 2009, the genomes of each were sequenced, opening the way to multiple studies including genetic studies concerning the virulence mechanisms of these bacteria and their very peculiar physiology.

bovis, a species that normally affects cattle. BCG does not prevent infection, but it prevents the tuberculosis infection from becoming deadly, particularly tuberculous meningitis and disseminated (or miliary) tuberculosis.

Vaccination with BCG was once mandatory in France, but now, as in many other countries where tuberculosis is not endemic, it is recommended only for so-called "at-risk" children. But who are these at-risk children in a developed country like France? This recommendation poses ethical problems due to its stigmatizing and discriminatory aspects, which the French *Comité Consultatif National d'Ethique* thoroughly discusses in its Avis 92 on tuberculosis screening and BCG vaccinations. In the United States, BCG has never been used for mass vaccination.

Childhood diseases

Pertussis and *Bordetella pertussis*

Pertussis, also known as whooping cough, is a respiratory tract infection that continues for several weeks following a week-long incubation period. It is characterized by coughing fits so violent that the patient struggles for air; in French, the disease is called *la coqueluche* because the patient's gasping sounds like the crowing of a rooster (*coq*). The disease's causative agent, *Bordetella pertussis*, was discovered by Jules Bordet and isolated in 1906. Previously characterized by a high mortality rate, this disease's prognosis has been decisively improved by antibiotics.

Vaccination has considerably reduced the incidence of pertussis. The original DTP combination vaccine (diphtheria, tetanus, and pertussis; used since the 1940s in the United States and since 1959 in France) protected against the effects of the disease but not against the initial infection. It has since been replaced by a new acellular version, more effective and better tolerated. Since vaccination became widespread after 1966, this newer

The diphtheria toxin

The mode of action of the diphtheria toxin has been studied in great detail. This toxin acts by modifying a very important protein of the host, translation elongation factor-2 (EF-2), thus blocking production of proteins in intoxicated cells, leading to cell death. In 1960, the diphtheria toxin was the first bacterial toxin shown to be capable of modifying a host protein posttranslationally.

acellular vaccine has been included in several combined mandatory vaccinations such as those against diphtheria, tetanus, poliomyelitis, and pertussis and against *Haemophilus influenzae* type b. Thanks to vaccination, pertussis has all but disappeared in young children in developed countries, but it is making a comeback in young adults, probably because immunity fades with time. It is strongly recommended that adults receive a booster vaccination.

Diphtheria and *Corynebacterium diphtheriae*

At the end of the 19th century, diphtheria was one of the leading causes of child mortality. It is caused by *Corynebacterium diphtheriae*, discovered by Edwin Klebs and Freidrich Löffler in 1884. The disease is characterized by the formation of false membranes at the entrance to the respiratory tract, causing suffocation and frequently death. Alexandre Yersin and Émile Roux showed that these clinical signs were caused by a toxin produced by the bacillus. The gene for this toxin is carried by a bacteriophage whose genome can integrate itself into and out of the bacterium's genome, producing either bacteria that can produce the toxin or entirely nonpathogenic bacteria. In 1890, Emil von Behring and Shibasaburō Kitasato, students of Robert Koch, discovered that patients cured of diphtheria carried "antitoxins" (antibodies) in their blood. This led to the idea of treating patients with "serotherapy" using serum obtained from recovered diphtheria patients or animals previously exposed to the toxin. In the 1920s, Gaston Ramon used inactivated toxins to develop the first effective vaccine. Diphtheria has now virtually disappeared.

Tetanus and *Clostridium tetani*

Tetanus is another acute disease caused by a single toxin, the tetanus toxin. *Clostridium tetani*, the etiologic agent in this disease—discovered by Shibasaburō Kitasato in 1889—is unique in that the bacterium produces

The tetanus toxin

The tetanus toxin, like the botulinum toxin, is a protease, an enzyme that can break down proteins. Both tetanus and botulinum toxins target SNARE proteins, membrane proteins that enable fusion between two membrane compartments, such as the fusion of an intracellular vesicle with a plasma membrane. By preventing vesicle fusion, tetanus toxin prevents neurons from releasing acetylcholine, a neurotransmitter.

spores that can remain dormant for years in the soil. Once these spores enter the host's body (through a cut, for example), they germinate in the anaerobic environment of the tissues and synthesize the toxin, which migrates to the central nervous system and causes the severe contractions and muscular spasms characteristic of the disease. Like diphtheria, tetanus can be prevented by vaccination with a chemically modified, nontoxic version of the toxin called *tetanus toxoid*. The disease is treated with antitoxins, sedation (with drugs such as benzodiazepines) to reduce the muscle spasms, and supportive care if the patient's respiration is impaired. Thanks to compulsory vaccination, tetanus has now nearly completely disappeared from industrialized countries; there are, for example, fewer than 10 tetanus deaths per year in France.

Streptococci

The *Streptococcus* genus comprises both pathogenic and nonpathogenic species. The three principal pathogenic species are (i) *Streptococcus pyogenes*, or group A streptococcus, which causes skin and lung infections and, most commonly, pharyngitis (sore throat) that can develop into rheumatic fever; (ii) *Streptococcus agalactiae* or group B streptococcus, causing neonatal, vaginal, and urinary tract infections; and (iii) *Streptococcus pneumoniae*, the pneumococcus, which results in ear and throat infections. Pneumococci also can cause fatal pneumonia.

S. pneumoniae is a common bacterium in the nose and pharynx—50% of the population carries it—and *S. agalactiae* is present in the vaginal microbiota of 30 to 50% of women. We do not know yet whether the commensal strains of *S. pneumoniae* can become pathogenic, or whether the pathogenic strains could become harmless commensals. Streptococci are usually transmitted by the saliva.

Streptococci as the basis for historic discoveries

During their development, pneumococci can change to become *competent*, that is, capable of taking up DNA that will change their properties. It was in pneumococci that Frederick Griffith's 1928 experiments first suggested the concept of transformation in bacteria. Further studies led to the discovery of DNA as the source of genetic information.

The group A streptococci are the bacteria in which tracrRNA, or trans-activating CRISPR RNA, was first detected. tracrRNA is coded upstream of the *cas* genes, which are themselves located upstream of a CRISPR element. tracrRNA guides the Cas9 protein to its DNA target (see chapter 4 and Fig. 10).

The first complete bacterial genome to be sequenced

In 1995, the genome of *H. influenzae* was the first bacterial genome to be sequenced in its entirety. It contains 1,830,140 base pairs and encodes 1,740 proteins.

Haemophilus influenzae

Described for the first time in 1892, this respiratory bacterium may be encapsulated or not. The nonencapsulated form is responsible for 40% of bacterial ear infections, while the encapsulated bacterium also causes ear infections but more often meningitis, septicemia, and pneumonia. The bacterium is often found in association with *S. pneumoniae* in respiratory infections. Antibiotic treatment is possible despite beta-lactamase resistance. Vaccines have been available since the early 1990s.

Meningococci and meningitis

Like streptococci, *Neisseria meningitidis*, also known as meningococcus, may reside in the human pharynx as a normal commensal; carriers can be perfectly healthy and asymptomatic. On average, *N. meningitidis* is present asymptomatically in 5 to 10% of the population, but this percentage can be as high as 50 to 75% in certain communities. Meningococci are passed from

Escaping the host's defenses and antigenic variation

N. meningitidis is characterized by its ability to greatly modify its surface. Surface proteins of a specific family may be replaced by a protein of the same family, slightly altered to avoid recognition by the host and specifically by the host's antibodies. This is "antigenic variation," a property of many pathogenic bacteria including *Neisseria gonorrhoeae*, the bacterial cause of gonorrhea. *Neisseria* spp. possess natural competence; they are able to internalize DNA at any phase of their growth.

human to human. Bacteria present in the nasopharynx can reach the bloodstream, cross the blood-brain barrier by escaping from cerebral microvessels, and ultimately target the meninges, causing meningitis. They can also multiply in blood and create a severe sepsis known as *purpura fulminans*. Meningitis and sepsis can be deadly, rapidly inducing death within a few hours if left untreated. Fortunately, *N. meningitidis* is still generally susceptible to antibiotics, so rapid diagnosis and treatment are crucial. Major neurological consequences may persist even after the patient has recovered, which is a further reason to attempt to reduce asymptomatic carriage of this bacterium.

Listeriosis and *Listeria*

Listeria was discovered in 1926 by E. G. D. Murray in England, during an outbreak in rabbits and guinea pigs in the Cambridge University animal care houses. It was not until later that it was identified as a foodborne pathogen in humans. *Listeria monocytogenes* is responsible for most cases of meningitis in newborns. Pregnant women are particularly susceptible to these bacteria, and it is assumed that many previously unexplained miscarriages were due to *Listeria* infection. *Listeria* is transmissible only through food; once the bacterium is in the intestine, it can pass through the intestinal barrier to reach its target organs, the placenta and the brain. Pregnant women are easily infected. Babies can be infected at birth, which is often premature. *Listeria* is opportunistic, infecting the elderly and immunocompromised individuals. It is therefore recommended that pregnant women and other individuals at risk avoid foods that may contain *Listeria*, including certain cheeses, raw milk, and prepared meat products. Food

Listeria, an invasive model bacterium

For the past 30 years, these opportunistic pathogenic bacteria have been studied by a combination of techniques in molecular biology, genetics, genomics, and cellular biology. *Listeria* has become one of the most extensively studied models for the biology of infections.

Listeria's virulence is due to its ability to resist the bactericidal properties of macrophages, to enter into nonphagocytic cells and multiply therein, and to cross three of the host's barriers: the intestinal barrier, the blood-brain barrier, and the placental barrier. The bacteria enter epithelial cells via surface proteins called *internalin proteins*, which interact with receptors present on the surface of mammalian cells (Fig. 16).

One of the most remarkable phenomena of *Listeria* infection is the bacterium's ability to move through cells and to transfer from one cell to the next by polymerizing cellular actin. Analysis of this mechanism—involving expression of ActA protein at one of the bacterial cell poles—led to the discovery of the first cellular actin nucleator, the Arp2/3 complex (see Fig. 18). Other recent discoveries made during studies of *Listeria* include several new types of regulation of bacterial gene expression by RNA regulators.

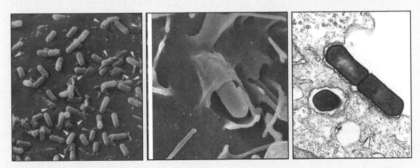

Figure 16. *Listeria* bacteria enter human cells. Images produced with a scanning electron microscope (at center and on the left) and transmission electron microscopy (on the right).

safety regulations target *Listeria* closely, which reduces the risk. Fortunately, *Listeria* is still susceptible to antibiotics, and therefore early diagnosis and treatment are critical to avoid meningitis and neurological sequelae. Listeriosis is also an important problem in livestock, primarily in cattle and sheep.

Intestinal infections

Cholera and *Vibrio cholerae*

Vibrio cholerae is responsible for highly contagious epidemics in humans, recognized since the 19th century in India and other Asian countries. It is generally considered that the bacterium was first isolated and identified as the agent responsible for cholera by Robert Koch in 1884, after a mission to Calcutta. However, there is some evidence that Filippo Pacini had isolated the bacterium in Italy in 1854. The infection is characterized by diarrhea, gastroenteritis, and vomiting that lead rapidly to dehydration that can be fatal if left untreated. The infection spreads by the fecal-oral route, generally by consuming contaminated water or food. It has a short incubation period of 2 hours to 5 days. The disease affects both children and adults and is prevalent in conditions with particularly poor hygiene and sanitation, conditions that can follow natural disasters, as occurred in Haiti after an earthquake in January 2010. The earthquake, with its toll of 220,000 fatalities, was followed by a cholera epidemic that cost over 8,500 more lives.

Cholera is easily treated with oral fluids and rehydration salts. In cases of severe dehydration, intravenous rehydration is used. In an epidemic, the highest priority should be access to an uncontaminated water source. Vaccines are available but have only a temporary effect.

Salmonellae: gastroenteritis and typhoid fever

Salmonella species can cause gastroenteritis and typhoid fever. *Salmonella enterica* serovar Typhimurium is responsible for foodborne gastroenteritis

---

Vibrio cholerae virulence factors

There are two principal virulence factors in *V. cholerae*: the cholera toxin that causes dehydration and the bacterial pili (type IV pili) that enable *Vibrio* to form biofilms. The toxin is carried by a lysogenic bacteriophage, whereas the pili, coregulated along with the toxin by the ToxR regulator, are encoded by a chromosomal pathogenicity island.

V. cholerae was one of the first bacteria found to have a type VI secretion system (see Fig. 14). This organism is also competent in certain conditions. It internalizes DNA by means of a highly sophisticated system that includes another pilus.

Virulence factors in salmonellosis

Salmonella virulence factors have been the subject of numerous studies. These intracellular bacteria live and multiply in macrophages but can also enter into epithelial cells. *S. enterica* has two type III secretion systems, the genes of which are concentrated in two pathogenicity islands, SPI-1 and SPI-2. The first allows the bacteria to enter cells and the second helps them to replicate in the internalizing vacuole. The proteins that the two systems secrete have a variety of functions. These include modification of the cytoskeleton of infected cells or inhibition of certain signaling pathways, particularly pathways involved in innate immune responses.

with symptoms including fever, diarrhea, vomiting, and abdominal pain. *Salmonella* contamination is found in a wide variety of foods. Healthy adults do not generally require specific treatment, but antibiotic therapy is recommended in the elderly, nursing mothers, or immunocompromised individuals who are at risk of serious infection or death. Various animals may also contract salmonellosis.

Typhoid fever is caused by *S. enterica* serovars Typhi and Paratyphi. Endemic in developing countries, typhoid is a serious illness that spreads via contaminated food or poor sanitation. It causes sepsis and high fevers that can develop into possibly fatal complications, though a small percentage of infected individuals can remain asymptomatic and unwittingly spread the disease through poor hygiene. Because *Salmonella* strains have developed antibiotic resistance, treatment options can be limited; currently the preferred drugs are fluoroquinolones and ceftriaxone. Typhoid can be prevented by vaccination, but the most important prevention strategy is improved sanitation and careful attention to food safety.

Escherichia coli and other coliform bacilli

Escherichia coli was discovered in 1885 by Theodor Escherich. This bacterium is called a *bacillus*, because its shape is elongated like a rod, to be distinguished from the spherical coccus form of the streptococci, gonococci, or staphylococci. *E. coli* is a major and usually beneficial component of our intestinal microbiota. There are, however, many pathogenic *E. coli* strains that can cause gastroenteritis, pyelonephritis (kidney infection), urinary tract infections, meningitis, sepsis, and septic shock. *E. coli* generally occupies the intestinal tract but may ascend to the bladder, the kidneys, and even the brain, depending on the individual species. Some of the most widely studied strains include uropathogenic *E. coli* (UPEC), enteropathogenic *E. coli* (EPEC), and other enterohemorrhagic *E. coli* (EHEC) strains, particularly the O157:H7 strain. O157:H7 is nicknamed the "hamburger bacteria" because of the severe outbreaks of this strain in Europe and the United States linked to the consumption of undercooked beef. EHEC strains produce the powerful Shiga toxin, which destroys intestinal cells.

In 2011, Europe was swept by a deadly outbreak of gastroenteritis (50 deaths) caused by EHEC strain O104:H4. At first blamed on contaminated cucumbers from Spain, the outbreak was later found to originate in salad sprouts grown from contaminated fenugreek seeds from Egypt. Intensive research has shown that this pathogen was an EHEC strain that had acquired genes from another category of *E. coli*, enteroaggregative *E. coli* (EAEC).

E. coli is easy to grow in laboratories, and ever since its discovery it has continued to be an excellent model for genetic as well as clinical research. François Jacob, André Lwoff, and Jacques Monod conducted their Nobel Prize-winning research in *E. coli*. The bacterium has been adapted by genetic engineering to produce industrial quantities of valuable proteins such as growth hormones or insulin, a process far safer and more efficient than the previous methods of extracting these proteins from human or animal cells. This process now avoids tragedies such as the development of prion-borne Creutzfeldt-Jakob disease in the 1980s and 1990s in people who had received growth hormones collected from human pituitary glands.

As with many bacteria, antibiotic-resistant *E. coli* strains have appeared, particularly with resistance to fluoroquinolones and cephalosporins.

EPEC and cellular adhesion: the astonishing story of Tir

E. coli strains vary greatly. Many are nonpathogenic, but others are armed with different virulence factors, particularly *pili*, that allow them to adhere strongly to epithelial surfaces. Some *E. coli* strains possess siderophores that allow them to absorb iron in iron-depleted locations such as the bladder, where UPEC forms biofilms before migrating to the kidneys. Certain strains produce toxins that can be very potent, such as the Shiga toxin in EHEC or the CNF1 toxin in UPEC.

EPEC possesses a unique strategy for colonizing the intestine. It uses a type III secretion system to inject Tir protein into the plasma membrane of the host's intestinal epithelial cells. Intimin, an attaching and effacing (A/E) protein expressed on the bacterium's surface, then binds to the embedded Tir protein. This intimin-Tir interaction alters the host cell's cytoskeleton, causing damage to the intestinal tissue that ultimately results in diarrhea, primarily in infants and children.

Health care-related infections

Enterococci

Enterococcus faecalis is an anaerobic bacterium that can be a harmless commensal in humans and other mammals but has the capacity to become pathogenic and cause urinary tract infections. The bacteria are resistant to numerous antibiotics and, consequently, are responsible for many hospital deaths due to infection.

Staphylococci

Like enterococci, the genus *Staphylococcus* contains pathogenic strains as well as nonpathogenic, commensal strains that belong to the cutaneous mucosal surfaces and can become pathogenic under certain conditions. *Staphylococcus aureus*, known as the *golden staphylococcus* because of the gold color of its colonies when grown on blood agar plates, is the most formidable of pathogenic staphylococci. It is perhaps the most common health care-related germ, and most of its strains have become multidrug resistant. These strains are responsible for multiple types of infections, including foodborne illnesses and various skin and mucous membrane infections.

The majority of specific symptoms of staphylococcal infections are caused by toxins. Cutaneous staphylococcal infections take many forms, from boils to abscesses. Some strains produce toxins specific to the genus called *exfoliatins* that affect the skin and cause it to peel dramatically, most notably in children (scalded skin syndrome). Infections of mucous membranes can rapidly evolve into sepsis. Staphylococci are responsible for most cases of toxic shock syndrome, which is rare but often fatal because of the enterotoxin produced. In hospitals, staphylococci often contaminate prostheses or implanted material. Antibiotic therapy is the recommended treatment, but strains resistant to vancomycin, methicillin, and other antibiotics have appeared, resulting in serious therapeutic impasses.

Virulence factors in enterococci

Enterococcus species are opportunistic. Factors involved in their virulence include proteins that help them to adhere to inert surfaces (such as medical tubing) or to cells as well as factors involved in the formation of biofilms. Enterococci also express a protease involved in the formation of biofilms.

Pseudomonas aeruginosa

P. aeruginosa has a relatively large genome (>6 million base pairs) and many virulence factors. It has been used for in-depth investigation of various compounds and phenomena including toxins (exotoxin A, exotoxin S), quorum sensing via homoserine lactones, type III and VI secretion systems, and type IV pili.

Pseudomonas: burns and cystic fibrosis

The genus *Pseudomonas* contains numerous species, both nonpathogenic and pathogenic, in humans, animals, and plants. These ubiquitous bacteria are found in a wide variety of places ranging from stagnant water to air heating and cooling systems. The most widespread and most studied species is *Pseudomonas aeruginosa*. *Pseudomonas* is an opportunistic pathogen that can cause secondary infections in burns and cuts. It is also responsible for a large number of health care-related infections because it is resistant to many chemical, antiseptic, and antibacterial agents.

P. aeruginosa is the most frequent pathogen affecting patients with cystic fibrosis (CF), a genetic disease affecting the lungs, among other organs. A CF patient's prognosis is directly linked to lung function and the avoidance of pulmonary infections. One of the many toxins *P. aeruginosa* produces is pyocyanin, a blue-colored metabolite that is particularly dangerous in patients with CF because it interferes with the function of cilia in the lungs. Antibiotic treatment against *Pseudomonas* in CF patients is of critical importance, even in the absence of direct signs of infection.

Some *Pseudomonas* species are capable of degrading certain chemical compounds such as hydrocarbons, and thus the species offers at least one potential beneficial purpose, in cleaning up environmental contamination, including oil spills.

Klebsiella species

Klebsiella species, particularly *Klebsiella pneumoniae*, are ubiquitous commensal Gram-negative bacteria present in the digestive tracts and respiratory systems of humans and animals. Like other commensal bacteria, they can become pathogenic under certain conditions. *K. pneumoniae* is the Gram-negative bacterium often responsible for health care-related pneumonia, as many strains have become resistant to beta-lactamases.

Sexually transmitted infections

Gonorrhea and *Neisseria gonorrhoeae*

N. gonorrhoeae, also known as the gonococcus, was discovered by Albert Neisser in 1879 and causes gonorrhea. The bacterium is sensitive to oxygen and drying and develops in the mucous membranes. Gonorrhea was once the most common genital bacterial infection but has been replaced in that dubious achievement by *Chlamydia*. The bacteria are only capable of spreading from human to human. In women, the infection is often imperceptible, but with particularly dramatic consequences including sterility. In men, infection may be quite painful, resulting in inflammatory urethritis that may cause chronic complications if left untreated. *N. gonorrhoeae* used to be easily treated with penicillin, but like many other bacteria, it has now developed resistance. Current treatments include ceftriaxone, azithromycin, and doxycycline, the last two of which are also used to treat *Chlamydia*.

Chlamydia trachomatis

Chlamydiae are "obligate intracellular" bacteria, meaning that it is not currently possible to grow them other than in mammalian cells. After human papillomavirus, herpesvirus, and *Trichomonas* infections, *Chlamydia trachomatis* infection is the most frequently reported sexually transmitted disease in the United States. The symptoms in both male and female patients are subtle and frequently overlooked. It is transmitted exclusively by human-to-human contact, either by unprotected intercourse or from mother to baby during birth. In men, *Chlamydia* can result in testicular and urethral infections; the effects in women are more serious, including pelvic inflammatory disease, problems with pregnancy, and even infertility. *C. trachomatis* also causes trachoma, a highly contagious infection of the eyelid that can result in blindness that primarily appears in regions without access to adequate medical treatment.

Diseases of armies

Typhus (not to be confused with typhoid fever, which is in no way related to it) decimated the armies of Napoleon during his retreat from Russia in 1812. It laid siege to the trenches in World War I and spread through Nazi concentration camps. For centuries, typhus epidemics have appeared during sieges and wars. Caused by the highly virulent bacterium *Rickettsia prowazekii*, the infection presents symptoms that include high fever, headaches, and extreme fatigue. It was named for Howard Ricketts and Stanislaus von Prowazek, zoologists who died while studying a typhus outbreak in a prisoner of war camp in 1915.

R. prowazekii spreads in conditions of poor sanitation, generally by body lice. When the louse takes a blood meal, the blood-borne bacteria multiply in the insect's gut. Feeding lice drop bacteria-laden excrement from which the bacteria can enter the human victim's skin through the bite wound or skin damaged by scratching. At the end of World War I, use of the recently discovered insecticide DDT (dichlorodiphenyltrichloroethane) helped to diminish typhus outbreaks. *R. prowazekii* is so dangerous and highly virulent that in the United States it is classified as a Select Agent pathogen, requiring a high-security facility and government clearance to work with it.

Agents of bioterrorism

Bacillus anthracis

Bacillus anthracis was first cultivated by Robert Koch in 1876, who discovered that it forms spores that can survive in the soil for years. These bacteria can infect both humans and a variety of animals (sheep, goats, etc.). In animals, it seems to spread via spores, either airborne or in food, and has a high mortality rate. Louis Pasteur, with the help of colleagues Émile Roux and Charles Chamberland, famously vaccinated sheep in Pouilly-le-Fort, France, in 1881 with a vaccine based on *B. anthracis* bacteria inactivated with potassium dichromate. This was one of the first examples of successful vaccination and made major headlines at the time.

In humans, *B. anthracis* infection takes three principal clinical forms. Cutaneous anthrax is acquired when spores from infected animals get into the skin through a scratch or other injury. With treatment, it is not serious. Gastrointestinal anthrax results from consuming spores in undercooked meat from an infected animal. It is rare in the United States, which has strict regulations about animal vaccination against anthrax. Inhalation anthrax is by far the most serious version. Acquired by inhalation of airborne spores, it can have a deceptively long incubation period that then leads to life-threatening sepsis. Inhalation anthrax has a near 90% fatality rate if untreated; with aggressive treatment, about 55% of patients survive. The five fatal anthrax infections in the United States in 2001, resulting from anthrax spores sent through the mail, were respiratory infection.

The tripartite protein toxin of *Bacillus anthracis*

B. anthracis has a plasmid that encodes three proteins, protective antigen (PA), edema factor (EF), and lethal factor (LF). These proteins form pairs of either PA and EF or PA and LF. PA binds to a cell and allows EF or LF to enter into the cell. EF is an adenylate cyclase that raises the concentration of cyclic AMP in cells, leading to major cell deterioration. LF is a protease that cleaves mitogen-activated protein kinase, leading to deleterious consequences including lysis of macrophages. The plasmid also codes for a capsule that surrounds the bacteria and prevents its ingestion (phagocytosis) by macrophages.

Helicobacter pylori

The transformation of urea into ammonia and carbon dioxide by urease allows *Helicobacter* to resist high acidity levels and colonize the stomach. These compounds are toxic and contribute to other factors expressed by the bacteria that produce the severe inflammation characteristic of *Helicobacter* infection.

New diseases

Helicobacter pylori

The role of *Helicobacter pylori*, named for its helical spiral shape, in gastric ulcers and gastritis was recognized thirty years ago. In 1982, Australian pathologists Barry J. Marshall and J. Robin Warren discovered the connection between gastric ulcers and *H. pylori* by the simple if drastic method of Warren's inoculating himself with the bacteria and promptly developing severe gastritis. It wasn't until 12 years later, however, that medical dogma finally acknowledged the link and antibiotics replaced antacids as the chosen treatment for gastric ulcers. Marshall and Warren won a belated Nobel Prize in 2005 for their studies.

H. pylori is an ancient bacterium long associated with humans, although it does not cause illness in most people. It is thought that *H. pylori* is carried by roughly two-thirds of the world's population. Genetic studies have shown evidence of the bacteria in *Homo sapiens* from 58,000 years ago! Left untreated, the ulcers can significantly increase the risk of gastric cancer; *Helicobacter* was one of the first bacteria clearly identified as being responsible for cancer.

Borrelia burgdorferi and Lyme disease

Like *Helicobacter, Borrelia* spp. are spiral-shaped bacteria. Although first discovered over a hundred years ago by Amédée Borrel, the species *Borrelia burgdorferi* was recognized only recently as the agent of Lyme disease. *Borrelia* species are spread by bites from infected ticks or lice carried by large undomesticated mammals such as deer, wild swine, and other animals. Lyme disease generally first appears as a red patch around the tick bite and later produces flu-like symptoms. It may result in muscle pain, neurological complications, and sometimes heart problems. Lyme disease is treated with antibiotics, although it is quite difficult to completely eradicate the bacteria if treatment is not begun promptly.

In addition to their atypical form, borreliae are characterized by the presence of multiple linear chromosomes, currently the subject of intensive research.

Legionella

Legionella pneumophila was discovered in 1977, following a mysterious outbreak of pneumonia that affected American Legion members attending a Bicentennial convention at the Bellevue-Stratford Hotel in Philadelphia in 1976. Of the more than 4,000 veteran soldiers present, 182 fell ill and 29 died. The infection was found to have spread through the hotel's air-conditioning system. Other outbreaks caused by *L. pneumophila* or similar *Legionella* species were later observed around the world, in France, Spain, Australia, Great Britain, the United States, and other countries. Legionellae reside in water and form colonies easily in cooling towers or water heaters that are not kept hot enough. It is now known that they can live and multiply in amoebae such as *Vermamoeba* (formerly *Hartmanella*) *vermiformis.* Even with antibiotic treatment, the mortality rate of patients with *Legionella* pneumonia is relatively high (10 to 15%).

Clostridium difficile

The *Clostridium* genus contains bacteria that are strictly anaerobic. *Clostridium difficile* takes its name from the difficulties Ivan Hall and Elizabeth O'Toole encountered when trying to isolate it in 1935. *C. difficile* is a normal commensal in the intestinal microbiota, but it is strongly opportunistic. Because it is resistant to most antibiotics, it tends to survive in intestinal microbiotas upon antibiotic treatment. It then overgrows the rest of the microbiota and causes gastrointestinal symptoms that are sometimes very serious. *C. difficile* is the main agent responsible for diarrhea in patients

Dot/Icm type IV secretion systems in *Legionella*

Legionellae are facultative intracellular bacteria; in their host, they multiply in the pulmonary alveolar macrophages. In these cells, the bacteria occupy vacuoles that form when the cellular membrane invaginates around the bacteria.

Legionella spp. possess a type IV secretion system, different from the type III secretion systems but equally complex. This system, known as Dot/Icm, allows the bacterium to generate over 100 proteins that have a variety of functions and to inject them directly into the infected cell.

Unexpectedly, the *Legionella*'s genome encodes a series of proteins that are normally found in eukaryotes, such as ubiquitin ligases. During infection these effectors are injected in the host cell.

undergoing antibiotic therapy, most notably in hospitals. It can form spores capable of surviving in many environments such as hospitals due to its resistance to typical disinfectants. Cases of "*C. diff*" infection are currently on the rise, in particular within elderly populations in assisted living factilities.

Diseases prevalent in developing countries

Clostridium botulinum

Clostridium botulinum, like *C. difficile*, is an anaerobic spore-forming bacterium found in soil. The spores are resistant to heat so that weak sterilization methods such as pasteurization (in which the temperature is raised to 70°C for a short period of time) may not effectively inactivate them. When spores germinate, they produce a toxin that can result in serious illness. Unlike *C. difficile* toxin, which causes muscle spasms, the botulinum toxin prevents muscle contractions, inducing generalized paralysis; if the respiratory muscles are affected, the victim can suffocate. Food containing this toxin, such as improperly sterilized canned goods or improperly heated or chilled prepared foods, can lead to severe cases of food poisoning. Today, however, this is fortunately quite rare because of increased public awareness of food safety principles. The botulinum toxin is used to treat certain illnesses such as improper eye opening and is also used for treating wrinkles—alas, only temporarily—by paralyzing the muscles responsible for them. This is the famous Botox.

Shigella flexneri as a model bacterium

Along with *Listeria* and *Salmonella*, *Shigella* is one of the most thoroughly studied pathogenic bacteria. Research has clarified the molecular and cellular bases for its virulence as well as its strategies for evading host defenses.

Shigellae are Gram-negative bacteria quite similar to *E. coli*. They possess a plasmid that carries many genes involved in pathogenesis, called the *virulence plasmid*. This plasmid includes genes for a type III secretion system and effectors that are transferred directly from the bacterium's interior to the eukaryotic cell during infection. For example, many proteins involved in *Shigella*'s entry into cells interact with the host cell's cytoskeleton, mimicking certain mammalian proteins; others display an enzymatic activity that interferes with the host's response to the infection. Studies on *Shigella* have shown that peptidoglycan plays a key role in inducing an innate immune response, which depends on intracellular receptors called Nod.

Shigellosis and diarrhea in the tropics

Shigellosis is a diarrheic illness that primarily occurs in tropical countries. It is responsible for hundreds of thousands of deaths per year, primarily in children younger than 5. It can be caused by several species of *Shigella*. *Shigella flexneri* is responsible for the endemic form of the illness. *Shigella dysenteriae* is responsible for the most brutal epidemics, and *Shigella sonnei* is sometimes responsible for infections in industrialized countries. Shigellosis is characterized by severe inflammation of the intestinal mucosa and is responsive to treatment with antibiotics. It is best prevented by improving conditions of hygiene. Vaccine trials are currently under way.

CHAPTER 13

The Multiple Strategies of Pathogenic Bacteria

Since the 1980s, pathogenic bacteria have been studied by a combination of novel approaches that show that they possess an incredible arsenal of infection strategies. After Louis Pasteur and Robert Koch, the study of bacterial pathogens consisted of isolating and classifying them, describing the infection by identifying which organs or tissues were infected, purifying and analyzing culture supernatants and their potential toxins, and then testing these in animals (mice or guinea pigs) or in cell cultures. Since those days, a wide array of new technologies has been developed, leading to a new era in infection biology.

Contributions of molecular and cellular biology

Following the emergence of molecular biology and genetic engineering in the 1980s, the study of pathogenic bacteria took off. Intensive research was spurred by the discovery of restriction enzymes—originally discovered as enzymes that bacteria produce to cut the DNA of bacteriophages that infect them—which won Werner Arber, Daniel Nathans, and Hamilton Smith a Nobel Prize in 1978.

At the end of the 1980s, researchers began isolating DNA fragments of bacterial pathogens and introducing them into minichromosomes called plasmids. The recombinant plasmids were then introduced into nonpathogenic bacteria to analyze their functions.

The second big step was the coupling of molecular biology approaches and cellular biology. Toward the end of the 1980s, investigators began to use mammalian cell cultures to analyze infections at the molecular level, in a somewhat reductionist manner, but nevertheless in a very precise way. At that time, optical microscopes and electronic microscopes capable of magnifying 10 million times were being improved. Confocal microscopes were first appearing, which use lasers to allow for high-resolution visualization of "marked" cells tagged with fluorescent compounds that recognize bacterial and/or cellular components. Better still, researchers started using video microscopy to observe bacteria in real time as they infected mammalian cells. Studies concerning the behavior of bacteria also allowed for new investigations into certain phenomena taking place in mammalian cells that had not been clearly understood until then. Bacteria themselves started to serve as tools for analyzing mammalian cells. This new field, combining microbiology and cellular biology, was named "cellular microbiology." Cellular microbiology showed that pathogenic bacteria deploy a range of powerful weapons that they use for successful infections while avoiding the host's defenses.

Bacteria that adhere to cells but do not enter them

All pathogenic bacteria have their own mechanisms of attack. Some do not enter their host cells; instead they adhere to the outside of the cell at a specific site. There they multiply and release their characteristic toxins that cause the host's illness. Some toxins kill the cell directly; others cause inflammation via lymphocytes and other immune-reactive cells attracted to the site whose sheer numbers can lead to tissue destruction. Among extracellular bacteria, some are poorly adherent but may produce toxins that disseminate throughout the organism and affect particular organs. The tetanus toxin from *Clostridium tetani*, for example, migrates to the central nervous system and disrupts synaptic transmission, causing seizures and paralysis. Some toxins, like the cholera toxin, cause severe efflux of water and ions from the intestinal cells they attack, resulting in watery diarrhea. Others, like the diphtheria toxin (see above), block protein synthesis by modifying translation machinery, leading to cell death.

The first cloning of virulence genes

The invasin protein in *Yersinia pseudotuberculosis*

Yersinia pseudotuberculosis was known to be able to enter into mammalian cells. To identify the invasion gene of *Y. pseudotuberculosis*, the bacteria's DNA was isolated, purified, and cut into pieces by restriction enzymes. The fragments were then inserted into plasmids that were introduced into the model bacterium *Escherichia coli*. All the transformed strains were tested for their capacity to enter into mammalian cells in culture. One of these resulting *E. coli* strains, transformed by a plasmid carrying a fragment of only 3 kilobase pairs (3,000 pairs of nucleotides), was able to penetrate into mammalian cells. This experiment showed that on a single fragment of the *Yersinia* chromosome there is a gene that codes for a protein enabling them to enter into cells. This protein is called *invasin*.

Invasion genes in *Salmonella*

It has long been known that *Salmonella* spp. are also capable of entering mammalian cells. As with the studies of *Y. pseudotuberculosis* above, *Salmonella* DNA was isolated, purified, and cut into large pieces by restriction enzymes, then inserted into plasmids and introduced into *E. coli*. One of the resulting strains, which had received a plasmid carrying a large 40-kilobase-pair fragment from *Salmonella*, was able to enter mammalian cells in culture. *E. coli* strains carrying smaller *Salmonella* fragments were not able to enter into the cells. This experiment showed that *Salmonella* carries a series of genes on a large fragment of its chromosome that enable it to enter into cells. This type of grouping of genes is now referred to as a "pathogenicity island."

It must be emphasized that these cloning experiments were made possible by the fact that *Salmonella* and *Yersinia* bacteria are fairly similar to *E. coli*, the bacterium so essential to the rise of molecular biology. Nevertheless, this kind of molecular cloning experiment is still being used, with varying degrees of difficulty, in studies of bacteria responsible for important diseases, contributing to major advances in the identification of virulence factors in these pathogens.

Toxins can act in very different ways and on a variety of targets, on cell membranes or inside cells. They can modify cellular components, reversibly or irreversibly, or even completely destroy them. They can take on the role of an inhibitor simply by sequestering one of the cell's compounds, or they can modify this compound by acting as enzymes.

Few bacteria cause illness due to a single toxin. Infection and disease are generally the result of a combination of factors. Enteropathogenic *E. coli* (EPEC), for example, injects the Tir protein into intestinal cells once it reaches the intestinal mucosa, where the protein acts to anchor the bacteria on the surface of the intestinal epithelium. EPEC then uses its type III secretion system like a sort of cannon to keep injecting a series of proteins that target different parts of the infected cell. Some of these targets are essential for maintaining the normal function of the intestinal villi. As the gastroenteritis provoked by EPEC develops, these villi disappear, throwing the intestinal barrier into chaos.

Many bacteria use the type III secretion system to inject proteins into their target cells, but the proteins they inject vary depending on the bacteria. The type III system of *Y. pseudotuberculosis* injects enzymes into macrophages—cells which are normally responsible for eradicating invaders—that prevent the macrophages from internalizing and destroying the yersiniae. These enzymes are defined as *antiphagocytic* proteins.

Salmonella and *Shigella* bacteria use their type III secretion systems to inject into cells proteins that attack the cells' structural compounds, their cytoskeleton, or their membrane and cause them to internalize the bacteria. Once safely inside the host cell, these bacteria continue secreting toxins, disrupting the cell's equilibrium and its interactions with its neighbors. In contrast, the role of proteins produced by the type III secretion system of *Chlamydia* spp. is still little understood (Fig. 17).

There are several other types of secretion systems similar to the type III secretion system that act like cellular cannons or nanomachines that

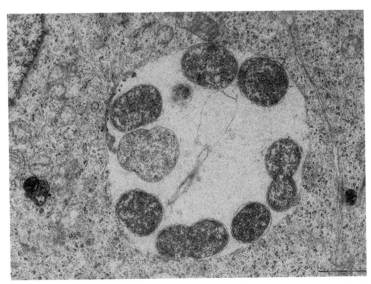

Figure 17. *Chlamydia* bacteria inside of a vacuole formed after entering human cells. Some bacteria are in the process of dividing.

translocate proteins into cells and hijack and/or disrupt host cell pathways and mechanisms; these include the type IV secretion system used by *Legionella* and the type VI secretion systems previously mentioned that are used in interbacterial combat.

Invasive bacteria

Some bacteria, like *Listeria monocytogenes*, are able to enter cells even though they lack a type III secretion system (Fig. 18). They actively multiply inside cells by hijacking all the nutrients that are present there. Their ability to enter into cells is mediated by two proteins, internalins A and B, present on the bacterial surface; these interact with receptor proteins on the surface of eukaryotic cells that usually have other functions. This pirating allows *Listeria* to cross several of the host's barriers: intestinal, placental, even hematoencephalic (blood-brain). The way *Listeria* crosses the intestinal barrier, to reach the bloodstream and travel through the body, is an interesting process. Although intestinal cells do not homogeneously express the internalin

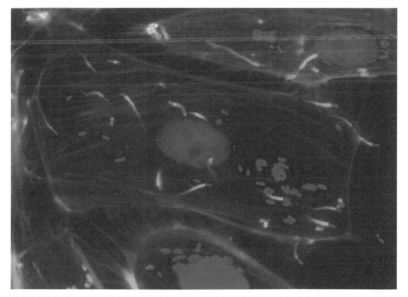

Figure 18. Cell infected with *Listeria* and actin comets. *Listeria* bacteria in mammalian cells recruit actin and polymerize it on its posterior pole. In this figure, the bacteria have been labeled with fluorescent anti-*Listeria* antibodies (red) and the actin (green) was detected with a specific marker (fluorescein isothiocyanate phalloidin). The nuclei of infected cells are labeled with DAPI (blue).

A receptor, the bacteria manage to find locations where it is available, for example at the tip of the intestinal villi. Intestinal epithelial cells have a life cycle in which they migrate from the base of the villi, where they are created, up to the tip, where they die by apoptosis and detach. This exposes the receptor E-cadherin, to which *Listeria* attaches with high affinity via internalin A. Another site of entry is the goblet cells, the cells that secrete the mucus. This provides the bacterium with an easy route through the interrupted mucous layer and epithelium and on into the bloodstream.

Listeriae are fascinating bacteria in part because once a bacterium gets inside a cell, it can recruit actin (a cellular protein partially responsible for the cell's plasticity) from the cell and form it into long filaments that extend from one of the bacterial poles. This creates a sort of actin "comet" that propels the bacterium with surprising power through the cell's internal membrane in order to pass from one cell to the next. This is how *Listeria* spreads from cell to cell sheltered from antibacterial compounds such as antibodies (Fig. 18).

Listeria's actin-dependent motility is remarkable. The bacteria move at the rate of approximately 10 μm per minute and are therefore able to cross a mammalian cell in 5 minutes. The analysis of this phenomenon provided insights into the mechanisms of how human cells move. When inside the host cell, *Listeria* expresses on its surface a protein called ActA that imitates WASp proteins of the cell whose functions were previously unknown. Using ActA, the bacterium takes over compounds involved in the mechanisms that control the cell's shape and movement. Thus, studying the ActA bacterial protein has given us a better understanding of fundamental cell mechanisms, which can cause cancer and metastasis when they are disrupted. ActA is one of many examples where studying an infection process has led to unexpected advances in cell biology.

Inside the infected host cells, bacteria can feed, multiply, and eventually pass from one cell to the next, all while being protected from the cell's mechanisms of defense. Among the "hijacking mechanisms" used by bacteria is a strategy in which they inject what are called *nucleomodulins* into the nuclei of infected cells. These proteins enter the cell nucleus and literally reprogram the cell.

Eukaryotic cell nuclei contain chromosomes (humans have 22 pairs of homologous chromosomes and one pair of sexual chromosomes, XX or XY, thus a total of 23 pairs) made up of cellular DNA and associated proteins. These form a dense complex, *chromatin*, which can relax under certain conditions. *Heterochromatin* occurs in highly compact regions of the chromosome, and *euchromatin* forms when the regions are more freely distributed, as this gives access to compounds and enzymes involved in transcription. Chromatin is compact if the compounds that form it, the nucleosomes, are close together. If the nucleosomes are more spread out, the chromatin is less compact. It has been found that some bacteria secrete proteins capable of entering into a cell's nucleus and altering the structure of chromatin, consequently altering the cell's transcription. These proteins are the nucleomodulins. Some nucleomodulins modify the structural compounds called *histones* that are essential to the structure of nucleosomes; others alter the structure as well as the transcription of chromatin complexes.

A major issue is, do chromatin modifications caused by pathogenic bacteria—referred to as *epigenetic* because the changes are not encoded in the genomic sequence—continue when the cell no longer contains the bacteria? If so, it would imply that cells can "remember" a past infection and, somehow, be preactivated in order to ward off identical or similar infections. This question is being investigated in several research laboratories.

The benefits of genomics

Study of the biology of infections has benefited as much as other disciplines, if not more, from advances in genomics—the study of genomes. The first bacterial genome to be completely sequenced was that of the pathogenic bacterium *Haemophilus influenzae*, responsible for bronchopulmonary and ear infections in children. It was sequenced in 1995 at the Institute for Genome Research founded by Craig Venter. Many other bacterial genomes have since been successfully sequenced. At present, sequencing of bacterial genomes takes far less time than when it first began (just a day, versus months or years) and is no longer a prohibitively expensive process (from several million euros to less than €1,000 or just over $1,000 per genome).

Advances in genomics have allowed *postgenomic* approaches to develop, which are completely different from approaches based on classic genetics. Instead of researching a single mutant bacterium, identifying and characterizing the mutation, and then trying to understand the molecular basis of the mutant's properties—the phenotype—researchers now study individual genes and determine their function by constructing a mutant and then analyzing the properties of the mutant. This approach is known as *reverse genetics.*

Postgenomic comparative studies are also very informative. Comparing the genome of a pathogenic bacterium with that of a similar bacterium that is not pathogenic allows the identification of genes potentially responsible for the infection. The *L. monocytogenes* genome, for example, was compared in this way with that of *Listeria innocua.* Sequencing techniques also make it possible to study all the minor variations between different strains of the same bacterium to understand why some are more virulent than others, or whether a bacterium's genome acquires variations during persistent infection. Many studies are now focusing on the persistence of infections.

In conclusion, many strategies used by pathogenic bacteria have been revealed by the use of a combination of novel techniques ranging from molecular biology to genomics and genetics, cellular biology, and many new imaging techniques. Although cellular microbiology has revolutionized our view of infections, it is clear that another revolution is taking place, one that will allow us to understand the role played by microbiotas present in different locations in the body in fighting infections, to identify the respective roles of various bacteria in a bacterial population. The next step will be to assess the validity of conclusions drawn from cell culture to the context of the whole body.

CHAPTER 14

Pathogenic Bacteria in Insects

Insects make up the most diverse form of animal life on the planet. There are close to a million catalogued species of insects in the world, which is more than all other types of animals put together. Insects can be infected by viruses, bacteria, fungi, and parasites, but little is yet known about the mechanisms of these infections. The first studies were motivated by economic factors. Pasteur's interest in silkworms in 1865 was inspired by a disease that was affecting the French silk industry. He discovered the disease was due to a microsporidium, a unicellular fungal parasite, in the silk worm's hemolymph that was "vertically" transmitted from generation to generation via infected germ cells.

More recently, diseases affecting bees have become problematic, particularly those that have an effect on the developing larvae in the hive (brood), including the European foulbrood caused by *Melissococcus pluton* bacteria and American foulbrood caused by *Paenibacillus larvae*. Nevertheless, we already know that colony collapse disorder in bee colonies is caused not entirely by bacteria, but by a combination of microbial and environmental factors.

It has been established for some time that insects can be carriers for various pathogens, inspiring much interest in the interactions between insects and microorganisms. Interestingly, it was long believed that insects were passive vectors that could transmit microbes on their surfaces or by regurgitation or defecation or by biting, but the situation appears to be much more complicated, varying according to the insect. We realize that we need more knowledge about insects and pathogens in order to better organize protection against infectious diseases potentially carried by insects. Much of the fundamental research on this subject has been conducted on the fruit fly *Drosophila melanogaster*. Bruno Lemaître and the group led by Jules Hoffmann conducted research on this fly that resulted in the discovery of general principles of *innate immunity*, earning Hoffman a Nobel Prize in 2011. Organisms with innate or natural immunity possess recognition mechanisms for certain compounds that they do not produce themselves but that are often found in pathogenic agents, such as the peptidoglycan found in bacteria. Recognition of these compounds stimulates nonspecific defense mechanisms as one of the host's first lines of defense. The normal healthy host's immune system will then trigger more specific defense mechanisms—an ability called *adaptive immunity* that is found in higher organisms but not in insects.

Insects are present in many different environments, where they interact with a variety of bacteria in relationships that range from symbiotic, mutually beneficial coexistence to drastic and fatal infections. *Drosophila* flies, for example, can be infected with bacteria such as *Erwinia carotovora, Pseudomonas entomophila*, or *Serratia marcescens*. Insects primarily seem to contract these bacteria by eating contaminated food.

Insects have in fact several lines of defense against bacteria. The first is their protective cuticle. Bacteria, however, can get past this via damage to the exoskeleton; they can also gain access to the insect's interior through the mouth, the anus, or the spiracles (openings along the insect body that serve the purpose of lungs). The bacteria *Photorhabdus* and *Xenorhabdus* spp. have been observed to infect insects by means of hitchhiking inside the intestine of a nematode or worm that burrows into the insect larvae. When the nematode reaches the insect's hemolymph, the bacteria are released and produce a range of toxins that kill the larva. The bacteria then feed on the carcass, producing nutrients for the nematode. Bacteria in insects can also be vertically transmitted through eggs or germ cells, as mentioned above for *Wolbachia*.

Although little is known about how pathogens kill insects, it is clear that the bacterium must counteract the insect's immune system in order to survive and proliferate. Certain bacteria, such *Bacillus thuringiensis*, produce specific toxins that destroy the insect's intestinal cells. Enzymes such as lipases, proteases, and hemolysins appear to play a role in infection. Additionally, some entomopathogenic bacteria produce toxic secondary metabolites.

To conclude, infections in insects, like infections in other animals, are the result of a combination of factors, including bacterial resistance to the host's initial defenses and a variety of factors specific to the bacteria. As insects can be highly beneficial—for example, for plant pollination or as food—a better understanding of insect physiology and their defense systems is needed.

CHAPTER 15

Plants and Their Pathogenic Bacteria

The world of plants has its own share of diseases, due only in part to bacteria. The primary pathogens in plants are fungi, which cause the majority of plant diseases.

Phytopathogenic bacteria—bacteria that damage plants—can enter plants at various locations ranging from natural openings to injuries caused by phytophagous insects or natural events. They can engender a variety of symptoms, including leaf spots of localized decay; soft spots of rot caused by the rapid proliferation of bacteria and resultant tissue destruction; tumors known as galls caused by the unregulated proliferation of certain cells; and wilting, which is generally due to bacterial proliferation inside plant tissue.

Diseases in plants have huge economic repercussions. Citrus canker disease, caused by *Xanthomonas citri* bacteria, led to the loss of millions of trees in the United States and Brazil. Pierce's disease in grapevines, caused by *Xylella fastidiosa*, has threatened the wine industry in certain regions of the United States. Another strain of *X. fastidiosa* is currently killing off olive trees in the Apulia region of southern Italy, with devastating consequences for the region's economy. This species, which does not appear to affect grapevines, is transmitted by the froghopper or spittlebug; the bacteria form biofilms in this insect's intestinal lumen and are regurgitated out onto the plant.

Most phytopathogenic bacteria belong to one of the following genera: *Acidovorax, Agrobacterium, Burkholderia, Clavibacter, Erwinia, Pantoea, Pectobacterium, Pseudomonas, Ralstonia, Streptomyces, Xanthomonas, Xylella, Phytoplasma,* and *Spiroplasma*. Some have been well studied (see below), while others deserve investigation.

Many plant pathogens use strategies quite similar to those found in animal and human pathogenic cells. For example, the type III secretion systems that allow *Yersinia, Salmonella,* and *Shigella* to inject effector proteins into host cells, then enter the cells themselves, are also used by bacteria like *Ralstonia solanacearum* to inject dozens of effectors into plant cells. Plants have defense systems that are to a certain extent analogous to those found in animals, although this is beyond the scope of the present discussion.

Agrobacterium tumefaciens and genetically modified organisms

Agrobacterium tumefaciens is, as its name indicates, a bacterium that induces tumors in plants, particularly in trees. It possesses a plasmid called the Ti plasmid. Once a bacterium makes contact with a plant cell, it can inject a fragment of this plasmid, T-DNA, into the nucleus of the plant cell, where it will integrate into the genome. The T-DNA is then expressed, leading to the synthesis of the plant growth hormones auxin and cytokinin, causing abnormal cell multiplication in the infected plant and forming a tumor. This also causes the plant to produce compounds called *opines*, which *Agrobacterium* uses via proteins encoded by Ti plasmid genes.

Agrobacterium and specifically the Ti plasmid have been the basis for numerous genetic manipulations, making certain cultivated plants resistant to nonselective herbicides or giving plants resistance to certain insects by inducing them to produce the *Bacillus thuringiensis* toxin. These plants endowed with foreign genes are called *genetically modified organisms*, or GMOs.

How can phytopathogenic bacteria be controlled? Chemical sprays, with copper as the active ingredient, can be used; certain bacteria, however, have become resistant to copper. Antibiotic use in plants is illegal in most European countries, with some exceptions, for example in the case of fire blight in Belgium. Fire blight, caused by the bacterium *Erwinia amylovora*, can be deadly for fruit trees, including apple, pear, and quince trees. To prevent the spread of this kind of plant disease, the recommended solution is to cut off all affected parts of the plant and burn them.

Phytoplasmas: pathogenic bacteria in plants and insects

Phytoplasmas are a type of small bacteria, pathogenic to plants, that lack a cell wall and somewhat resemble the mycoplasmas that are pathogenic to humans and animals. They are obligate symbionts in insects, never found living freely in nature, and are transmitted by the insects they depend upon. Phytoplasmas have been responsible for severe crop losses around the world. These bacteria multiply intracellularly in plants and insects, but they can also multiply extracellularly in insects. In insects, phytoplasmas invade the intestine and cross it to reach the salivary glands. When the infected insect feeds, the bacteria are introduced into a plant's phloem (the tissue that carries sap). Symptoms usually appear in infected plants after a week but can sometimes take much longer (6 to 24 months), depending on the strain of phytoplasma and the species of plant. Due to their long incubation period in plants and insects, phytoplasma epidemics are often detected far too late, for example just before harvest, which helps the bacteria spread. When an uninfected insect feeds from the phloem of a contaminated plant, it can spread the infection from 7 to up to 80 days after taking its infected meal.

Phytoplasmas have a broad spectrum of insect hosts. A particularly well-studied phytoplasma is that responsible for aster yellows, spread by aster leafhoppers. Phytoplasmas secrete several proteins that diffuse through the plant with some that can reach cell nuclei. There, they interact with transcription factors to inhibit the production of the plant hormone jasmonate, weakening the host's defenses and providing their insect vectors with favorable conditions for laying eggs. There are other effectors of plant bacteria that go to the nucleus, such as the TALE (transcription activator-like effector) proteins, the effectors of *Xanthomonas* that cause plant yellowing and wilting. In the plant *Arabidopsis*—a small plant in the mustard family with a very short life cycle that is a favored model system in labs—infected flowers have green petals.

Phytoplasmas cause other symptoms that suggest they interfere with plant development. Typical symptoms include "witch's broom," in which many branches grow all grouped together, and phyllody, when leaves grow in place of flowers, stalks redden, and the phloem decays.

The genomes of two phytoplasmas have been sequenced. They are small, like those of other endosymbionts in insects; phytoplasmas have lost many other genes as well, specifically those involved in forming cell walls.

While the phytoplasmas are pathogenic for plants, they may either affect the insects that carry them or be asymptomatic, depending on the phytoplasma and on the insect. For example, they increase the insects' fertility rates and can affect their flight patterns and their preference for one plant and not another. They can also manipulate plants into becoming the hosts for other insects.

Insect phytoplasma vectors are sensitive to the cold, which suggests that global climate change could enable a greater proliferation of these insects and their bacteria. There is therefore the risk of an increase in phytoplasma infections in the years to come.

Growing interest in organic food and agriculture stimulates research on plants, their diseases, and the vectors for these diseases, which are primarily insects.

CHAPTER 16

New Visions in Infection Defense

Genetic theory of infectious diseases

Only some individuals become ill when exposed to pathogenic bacteria or other microorganisms, and the severity of the illness can vary from person to person. This may be due to variations in the pathogen responsible for the illness (some strains of a bacterial species can be more virulent than others), to environmental factors, or to genetic differences between infected individuals.

If infectious diseases are unquestionably due to pathogenic agents, it is also clear that not all individuals are equally susceptible to these agents and that the "terrain," as Pasteur mentioned it—that is to say, the clinical backdrop or landscape—plays a significant role. Charles Nicolle, the noted author of *Naissance, vie et mort des maladies infectieuses* (*Birth, Life and Death of Infectious Diseases*) and 1928 Nobelist, first defined asymptomatic infection. The big question is, what are the fundamentals of clinical heterogeneity in infected populations?

The genetic theory of infectious diseases proposes that genetic factors determine the predisposition or resistance of a population to infectious disease. From the 1920s to the 1950s, research in genetic epidemiology laid the foundation for this theory by indicating there could be a determining link between disease and genetic predisposition.

On the molecular level, the complex genetic link to disease (in this case, a common infectious disease plus multiple predisposition genes in the victim) was acknowledged in 1954 with the discovery that sickle cell disease provided resistance to malaria. Sickle cell disease, or sickle cell anemia, results from a mutation in a gene that encodes one part of hemoglobin. This causes red blood cells to become sickle shaped, which impairs the reproduction of the *Plasmodium* parasite responsible for malaria, thus providing resistance to malaria infection.

A Mendelian approach to the genetics of infectious diseases (here a rare monogenetic defect in the patient that affects several infectious diseases) developed in 1952 with the discovery of X-linked agammaglobulinemia, which causes susceptibility to bacterial infections in the respiratory and digestive tracts of children. This immunodeficiency disorder, linked to the X chromosome, results from a genetic mutation in the gene encoding Bruton's tyrosine kinase (BTK). BTK is involved in the maturation of B lymphocytes, so this mutation prevents the development of healthy B cells. This finding thus showed that certain human genes can affect vulnerability or resistance to specific infectious diseases.

Spurred by this discovery, a research group led by Jean-Laurent Casanova identified more than six genes involved in BCGitis, infections resulting from the BCG vaccine. Children infected with BCGitis were shown to be immune to other infectious agents, with the exception of salmonellas. The first mutations identified were the genes involved in immunity, particularly in the signaling pathway crucial to defending against mycobacteria and salmonellae: the interleukin-12/interferon-γ pathway.

Additional studies on other infectious diseases in children, such as invasive pneumococcal infections and herpetic encephalitis, have demonstrated the importance of genetic variation. As in bacteria, even small genetic changes can greatly impact infections and even a single mutation can have dramatic consequences.

Health security in the age of globalizing risks

The "One Health Initiative" by the World Health Organization (WHO) and by the World Organisation for Animal Health (OIE) was born from increasing awareness, over the course of recent epidemics, of the health risks posed by globalization and of the role played by interactions between humans, animals, and environments in the emergence of pathogens. It has highlighted the need for research collaboration between the fields of human health, animal health, and environmental resource management, calling for physicians, veterinarians, and environmental scientists to work together.

Roughly 60% of known infectious diseases and 75% of emerging diseases in humans can be traced to animals (zoonoses). The same can be said for 80% of pathogenic agents that have the potential for use as tools of bioterrorism. The numerous possibilities for interspecies transmission, be it through food, the air, or simple contact, require solutions born from collaboration between sectors of human and animal health.

Migrations, changes in ecosystems brought about by deforestation and urbanization, and climate change are contributing to the emergence of new diseases. Many pathogens prove themselves capable of posing a risk to humans when their natural environment is disturbed. Biodiversity therefore constitutes a protective barrier against diseases, specifically those transmitted by vectors, as disturbances in ecosystems can engender the proliferation of species that carry pathogens or even introduce these pathogens in humans.

Changes in production methods and livestock-raising practices provide favorable conditions for the circulation of pathogenic agents, microorganisms and their vectors adapt and develop resistance. Highly contagious animal diseases can have drastic economic as well as human consequences. Similarly, plant pests can negatively affect food security and public health due to declining agricultural production as well as the presence of toxins or allergens. Preserving ecosystems and understanding animal and crop pathogens are both vital to global food security.

The observations presented here demonstrate the importance of supporting increased collaboration among research in human and animal health, studies on food security, and environmental surveillance on an international scale. These priorities link the concept of "Global Health," a recent initiative seeking to globalize health for all individuals regardless of circumstances, with the One Health Initiative, which seeks a collaborative, holistic surveillance of the health of the environment, animals, and people.

PART IV

Bacteria as Tools

CHAPTER 17

Bacteria as Tools for Research

Fundamental research on bacteria by searching to characterize their specific properties, the mechanisms they use for survival, and for maximally using resources available in the various niches where they reside and multiply has yielded sometimes astonishing and unexpected developments. From the discovery of penicillin to the genomic modification and genome-editing techniques based on the CRISPR/Cas9 system (CRISPR is clustered regularly interspaced short palindromic repeats), basic research has unveiled countless principles that have become the foundation for breakthroughs, subsequent studies, and new applications. This chapter highlights some of the most significant and useful discoveries over the years.

Restriction enzymes

Bacteria are able to protect themselves from the viruses known as *bacteriophages* by using the CRISPR system to defend themselves after the first encounter with a specific phage. If the strain has no CRISPR mechanism, however, or has one but has not encountered that particular phage before, the bacteria have another defense system: restriction enzymes. As soon as the bacteriophage injects its DNA into the bacterium, these proteins cut the DNA at specific recognition sites and deactivate it. To protect themselves from their own enzymes, bacteria use modification enzymes to alter their own DNA at the same sites.

Each bacterial species cleaves DNA from invading phages at a particular sequence. For example, the *Escherichia coli* RY13 strain contains a plasmid encoding the EcoRI restriction enzyme, which recognizes a specific cleavage site sequence in DNA (GAATTC). Because this sequence happens to be fairly common in bacteriophage DNA, the EcoRI enzyme can cleave the invader phage's DNA to protect the bacterial cell against infection. In order to protect its own DNA from a similar fate, the bacterium uses a methylase enzyme to modify the second A at any place in its own DNA that the cleavage site sequence appears. When this methylation modification is written out, it is shown as GAA*TTC. The methylation changes the cleavage site sequence in such a way that the EcoRI enzyme cannot cleave it, thereby protecting the bacterial DNA.

Other examples of restriction enzymes are found in *Haemophilus influenzae*, which produces the enzyme HindIII, and in *Bacillus amyloliquefaciens*, whose restriction enzyme, BamHI, cuts DNA at the sequence GGATCC. Of course, restriction modification systems are not 100% effective, which explains why viruses are still able to kill many of them. Bacteria also depend on CRISPR systems, described above and in chapter 4, that allow bacteria to retain memory of phages they encounter so as to defend themselves better if they encounter the same phage a second time.

When first investigating restriction enzymes, researchers had to purify the enzymes from bacterial cultures. Today, restriction enzymes are commercialized, with many companies invested in this lucrative enterprise. Several hundred restriction enzymes are now available that have a wide range of recognition sequences and function in highly diverse conditions. Enzymes from thermophilic bacteria (such as bacteria found in hot water sources like the geysers in Yellowstone National Park) are able to cleave double-stranded DNA at very high temperatures. Restriction enzymes are routinely used in the laboratory to isolate, clone, and analyze bacterial and viral genes. They are also used to clone eukaryotic genes in order to express them in bacteria to produce highly sought-after proteins such as hormones (particularly growth hormones), insulin, and others for a wide range of uses in medicine and research.

PCR

PCR, the polymerase chain reaction, is a technique used to detect and amplify DNA fragments from very small samples of DNA. It is used clinically to identify the presence of viruses, bacteria, or parasites in patients' blood samples or cerebrospinal fluid. Other applications include detecting traces of DNA that indicate the presence of microorganisms in food, cultures, and other materials. Forensic science now depends strongly on PCR methods, using DNA testing to change or even save lives by helping to establish innocence or guilt.

The PCR technique

The principles for PCR were first developed in the late 1980s by Kary Mullis, who received the Nobel Prize in Chemistry in 1993. The technique is based on the use of small fragments of DNA, called *primers*, that attach to complementary regions at opposite ends of a double-stranded DNA sample in a process called *hybridization*. The sample to be analyzed is first heated in order to split each double-stranded DNA fragment into two single strands. The sample is then incubated with the primers at a reduced temperature that promotes hybridization of the primers to the separated sample DNA strands. Then a bacterial enzyme capable of functioning at high temperatures is added; it recognizes the primers and synthesizes a complementary strand for each separated DNA fragment, resulting in a double strand for each original single strand. This process can be repeated for multiple rounds by sequentially increasing the heat to separate the strands, then lowering the heat to hybridize the strands and primers. After this cycle is repeated 30 times or so, more than 2 billion DNA fragments can be generated, enough that one can visualize them easily on an agarose gel.

The enzyme that extends the primer-bound DNA must be thermostable. This is why enzymes from thermophilic bacteria are used, such as the *Taq* polymerase found in *Thermus aquaticus*, a bacterium that thrives and multiplies at high temperatures. The technique above requires prior knowledge of the sample DNA sequence in order to use the appropriate primer. This is possible when searching a known pathogenic microbe, but what about when trying to define an unknown or emerging pathogen or the DNA of an unknown crime suspect?

Different variations in PCR methods have solved this problem by using general primers complementary to sequences that are very common in certain types of DNA. These types of primers are used when looking for bacteria in human body fluids or fecal matter or when studying microbiotas. Universal primers are then used that correspond to conserved regions in all bacteria, the sequences coding for 16S ribosomal RNA. Another alternative when searching for an unknown pathogen or organism is to use ligase—an enzyme present in bacteria and phages—to attach a primer to the DNA under study. Another primer can be hybridized on this primer. The next step is to proceed to the elongation and amplification steps as in the classic PCR described above.

Yet another improvement to the PCR technique is the technique of reverse transcription PCR (RT-PCR), which evaluates the amount of RNA present in a sample. The first step in RT-PCR is reverse transcription, in which the RNA is reverse transcribed into DNA that can then be amplified. RT-PCR is at the heart of the increasingly popular technique of RNA sequencing, which analyzes RNA produced in bacterial, mammalian, and other cells.

The PCR technique technology has become very widely used, and it is possible to amplify DNA sequences at highly accelerated rates (more than a billion times in under an hour).

Since the 1990s, the PCR technique has led to the emergence of the new disciplines of "paleomicrobiology" or "archeomicrobiology," in which investigators detect bacterial DNA in human skeletons and other remains. Several studies have focused on detecting the causes of major historical diseases such as tuberculosis (*Mycobacterium tuberculosis*) and the plague (*Yersinia pestis*). These studies have made it possible, for example, to identify *Y. pestis* unquestionably as the bacterium responsible for the Justinian Plague in the 6th century AD.

Bacteria and optogenetics

Various prokaryotes—archaea or bacteria, especially marine bacteria—harvest energy from light and use it in a variety of survival mechanisms such as phototaxis. Phototaxis is the regulation of movement by light: positive if it results in attraction (the organism moves toward a light source) and negative if it causes repulsion. Opsins are proteins that function in light collection. Type I opsins are present in prokaryotes and function in photosynthesis and phototaxis, while type II opsins—unrelated though similar in structure—function in the vision systems of animals. One category of the bacterial type I opsins in bacteria is called bacteriorhodopsin.

Even though animal and bacterial opsins are recognized as quite different proteins, their structures are similar. Studies of how bacterial opsins function, in particular bacteriorhodopsin, have led to wider investigations of opsins in other cell types that normally are not photosensitive, such as the

The rhodopsin family

Rhodopsins are compact molecules with seven transmembrane domains that use the cofactor retinal, a compound structurally similar to vitamin A that can absorb photons. When retinal absorbs photons, it changes form and opens the canal formed by the opsin's seven transmembrane domains, enabling ion transportation, such as proton export, or interactions with signaling proteins.

neurons in animal brains. The new technology of *optogenetics* studies the function of light to control biochemical mechanisms in eukaryotic cells. The first step is to introduce opsin into the subpopulation of neurons under investigation, which is done by viruses; alternatively, researchers may use transgenic mice that express rhodopsin via promoters that are specifically activated in neurons. Rhodopsins respond very quickly (within a millisecond) to light wavelengths tolerated by neurons. If the brain is exposed to light by means of optical fibers or other focused-light instruments, only the neurons that express rhodopsin react. In this way, the target neuron population can be distinguished and observed.

The optogenetics technique has led to major advances, including investigation of the role certain brain cells play in depression or addiction. This technique could even be applied to cells in other parts of the body, permitting a greater understanding of a wide range of physiological phenomena. Thus, the understanding of how bacteria react to light has given rise to valuable studies in much larger organisms, although the rhodopsin that is used routinely in optogenetic studies is not of bacterial origin.

The CRISPR/Cas9 revolution

Why does genomic modification/genome editing using CRISPR/Cas9 technology constitute a revolution? First of all, this technology is quick, easy to use, inexpensive, and effective. Moreover, it is based on a principle that is universal to all organisms. A guide RNA—a composite carrying a sequence identical to the DNA sequence to be studied—is introduced into a cell along with the Cas9 nuclease. The guide RNA detects the desired gene and guides the Cas9 to it, where the nuclease cleaves both strands of the DNA. The Cas9 protein can generate both deletion and insertion mutations.

This amazing new genomic modification technology can be applied to many fields ranging from fundamental biology to biotechnology to medicine.

Many applications are possible, such as the activation or visualization of genes. Researchers have even created genetic mutations or epigenetic variations, known to be associated with physiological defects or illnesses, within cells or animal models. Precise modification of plants such as wheat could confer resistance to adverse environmental conditions or infections, augmenting food security without the risk of introducing foreign DNA.

One might even envision the production of new biofuels by creating new metabolic pathways in algae or corn to alter them to produce ethanol. Gene therapy could be developed to target somatic tissues affected by genetic mutations or epigenetic variations. Finally, CRISPR/Cas9 technology could generate bacteria capable of synthesizing medicines or their precursors on a massive scale.

Numerous start-up businesses have pounced on the potential applications of this technology as a research tool or even for the manufacture of complete cell lines, bacterial mutants, or genetically modified animals. Earlier techniques based on meganucleases, zinc-finger nucleases, or TALE proteins (originating from the plant bacterium *Xanthomonas campestris*) are rapidly being abandoned in favor of the simpler and more rapid CRISPR technology. CRISPR methods are now being applied in diverse systems, from bacteria, mice, rats, rabbits, frogs, zebrafish, silkworms, *Drosophila*, and human cells to rice, sorghum, wheat, tobacco, watercress, and yeast.

Research now focuses on the most effective ways of introducing nucleases and their guide RNAs and on how to avoid introducing mutations in undesired locations. Inactive Cas nucleases, which are directed toward a genomic site by guide RNA but do not cleave the DNA, are also very attractive and powerful because they offer many possibilities, for example, high-precision visualization of chromosomal regions or sites.

Using pathogenic bacteria to understand eukaryotic cells

For millions of years, both pathogenic and nonpathogenic "commensal" bacteria have evolved with their hosts and adapted to the host's cells, developing strategies to protect themselves in order to survive the host's defenses. Studies over the past 30 years have shown that pathogenic bacteria are excellent cell biologists. They have evolved ways to enter cells, to recruit certain proteins in order to remain sheltered in replicative vacuoles, and to modify certain host proteins in order to survive inside the cell. Studies of these mechanisms have led to scientific advances in our understanding of fundamental mechanisms, three of which will be discussed in this section.

The ActA protein and cellular motility

A fundamental question in cell biology is to understand the mechanisms underlying cellular plasticity and motility, properties that are the foundation for many common normal phenomena, such as development in higher organisms or the response to infectious agents, as well as abnormal events such as the migration of cancer cells. How do cells organize themselves into an embryo? How does a signal from bacteria, for example in a skin injury, attract white blood cells (neutrophils) to an infection site? These questions were poorly understood at the end of the 1980s. It was known that cells polymerized long actin filaments in order for cells to change form, to leave their current location, and to move about. But how the first steps of actin filament formation occur was unknown. Studies of the pathogenic bacterium *Listeria monocytogenes* have played a key role in deciphering this phenomenon.

Discovering the role of the Arp2/3 complex

Listeria is an enteropathogenic bacterium normally found in the environment that can contaminate food. Once consumed, the bacteria can migrate from the intestine to more distant locations, such as the placenta and the brain. They travel via the bloodstream but also by means of the distinctive ability of this bacterium to enter its host's cells, where it can multiply and transfer easily from one cell to another.

Listeria capitalizes on the fact that all eukaryotic cells have a protein called actin that can assemble into long filaments and disassemble again. Our research group has shown that *Listeria* carries a surface protein called ActA that recruits a complex formed of seven proteins, the Arp2/3 protein complex. The Arp2/3 complex attaches itself to preexisting actin filaments in the cell, assembling them into relatively rigid filaments that propel *Listeria* into the cell's cytoplasm (see Fig. 18).

Studies of ActA led to the discovery of the Arp2/3 complex, without which bacteria would be incapable of moving inside cells. In eukaryotic cells, molecules capable of binding and activating the Arp2/3 complex form the family known as WASP/N-WASP; the proteins of this family are fairly close in structure to ActA protein.

The role of ActA and Arp2/3 in actin polymerization has been the object of intensive research for a number of years. Other nucleators, such as formins, have since been discovered.

Bacterial toxins

Many pathogenic bacteria secrete toxins that are responsible for the symptoms of the major diseases they cause. *Vibrio cholerae*, for example, produces a toxin in the intestine that is responsible for the major characteristics of cholera. *Clostridium tetani* and *Clostridium botulinum* produce neurotoxins responsible for the paralysis associated with tetanus and botulism. The tetanus and botulinum toxins are proteases, enzymes that cut and deactivate proteins; specifically, these two toxins damage the SNARE proteins involved in the transport of neurotransmitters. They have been quite useful in cell biology, especially so before the discovery of small interfering RNA (siRNA) techniques for deactivating cell proteins.

The C3 toxin in *C. botulinum* has been particularly useful for studying actin and the cytoskeleton in eukaryotic cells. C3 deactivates Rho, one of many "small GTPases" in cells, by adding ADP-ribosyl. Because Rho protein is involved in cellular plasticity, use of the C3 toxin has helped to shed light on the role of Rho in relation to other G proteins in different processes involving actin.

Nucleomodulins

Our research group has proposed the term *nucleomodulins* for the group of bacterial proteins that enter the nucleus of eukaryotic cells, where they interact with compounds normally involved in important functions, including DNA replication, chromatin remodeling, and DNA transcription or repair. Studies of these proteins have allowed remarkable advances in the understanding of fundamental mechanisms. Others, in particular the proteins injected by the plant pathogen *Xanthomonas*, have led to important applications.

The first nucleomodulins were identified in bacteria that cause disease in plants. One of the phytopathogens best known for injecting proteins and DNA into the nucleus is *Agrobacterium tumefaciens*. *Agrobacterium* injects its complexed (surrounded by proteins) T-DNA into plants, where it enters the cell nucleus and inserts itself into the plant's genome. This mechanism has not only allowed for greater understanding of regulatory phenomena in plants, it has also made it possible to generate transgenic plants. Genetically modified plants can express genes for resistance to general herbicides, which allows for increased yields of foods such as corn and sorghum, or other genes that encode the bacterial toxin of *Bacillus thuringiensis*, providing a defense against damaging insects such as the corn borer. Phytopathogenic bacteria are also the source of TALEN (for *t*ranscription *a*ctivator-*l*ike *e*ffector *n*uclease), a powerful research tool.

The TALEN technique

Xanthomonas campestris bacteria inject into plant cell effectors that act as transcription factors, or TALE factors (for transcription activator-like effectors), which attach themselves to specific sequences (TAL DNA box) by using a DNA recognition domain made up of repetitions of 34 amino acids. TAL domains, fused to the FokI endonuclease, are the basis for the efficient technique called TALEN, which allows for the modification of genes in plants as well as animals. Despite many improvements and much investment since its development just a few years ago, the TALEN technique has been surpassed by the simpler CRISPR technique.

Nucleomodulins have enabled the identification of some proteins that researchers had until now been unable to detect. For example, the nucleomodulin LntA produced by *Listeria* spp. interacts with a previously unidentified protein, BAHD1. This protein is part of a complex involved in heterochromatin formation and in the suppression of gene expression in mammals. LntA binds BAHD1 and removes it from the genes it targets, thus allowing those genes to be expressed.

The list of nucleomodulins keeps growing with the increase in studies in pathogenic bacteria.

CHAPTER 18

Bacteria: Old and New Health Tools

Over a century ago, Élie Metchnikoff proposed that human health could be improved and senility could be reduced by manipulating the intestinal microbiota with the beneficial bacteria found in yogurt. Metchnikoff received a Nobel Prize in 1908 for his discoveries on phagocytosis, the ability of white blood cells to engulf and deactivate pathogenic agents such as bacteria, viruses, and parasites. Although Metchnikoff could not have imagined everything we now know about the intestinal microbiota and its variations depending on age or diet, his ideas were visionary.

Bacteria in food

It is generally understood that yogurt, a fermented dairy product containing living bacteria that hydrolyze lactose, is good for your health. The two species of lactose-hydrolyzing bacteria usually used to make yogurt are *Lactobacillus delbrueckii* subsp. *bulgaricus* and *Streptococcus salivarius* subsp. *thermophilus*. As recognized by entities such as

135

the World Health Organization (WHO) and the Food and Agriculture Organization of the United Nations (FAO), as well as family doctors and individuals, eating yogurt can reduce the symptoms of lactose intolerance. But does yogurt contribute anything else to human health? Can yogurt bacteria be called probiotics? Well, yes and no. Evidence suggests that yogurt provides calcium ions and relieves symptoms linked to poor lactose digestion; it is also possible that individual bacteria involved in yogurt production can provide further health benefits. However, it is well known that the bacteria used to produce yogurt are not naturally present in the intestinal microbiota and cannot in fact survive in the gastrointestinal tract, whereas "true" probiotics can (see below). Several studies have been conducted to identify the health benefits of various bacterial strains in yogurt. One notable study showed that the capsular polysaccharide present in one strain of *Lactobacillus bulgaricus* stimulates a certain immune response in mice, in contrast to another strain of *L. bulgaricus* that does not have this polysaccharide. A subsequent clinical trial in humans indicated that the first cited strain could protect elderly subjects from rhinitis while the other strain tested did not. Another possible health benefit is that certain strains of yogurt bacteria can produce vitamins such as thiamine in the gut. Present knowledge seems to indicate that beneficial effects appear specific to particular strains.

A factor to consider is the synergy between strains used to ferment yogurt, as demonstrated in the case of aroma. It has been shown that yogurt's distinctive aroma is caused by the compound dimethyl trisulfide. This compound is present in small quantities when *L. delbrueckii* subsp. *bulgaricus* or *S. salivarius* subsp. *thermophilus* grow in monocultures but is produced in significant amounts by mixed cultures of *Lactobacillus/Streptococcus*.

Many other food products, mainly dairy products, result from bacterial fermentations, and the use of several bacterial strains in combination in certain food fermentation processes can lead to major advances in industrial biopreservation. An example is the combined activities of lactic bacteria and propionobacteria. Used in making Gruyère cheeses, this team of microbes has been found to have antimicrobial properties stemming from their production of lactic, propionic, and acetic acids, peroxides, diacetyl, and several other metabolites, plus bacteriocins as well.

Probiotics

In 2001, WHO and FAO defined probiotics as "live microorganisms which when administered in adequate amounts confer a health benefit on the host." Notable among the various microorganisms used as probiotics are the lactic acid bacteria, which are natural members of the human intestinal microbiota. The most studied probiotics belong to the genera *Bifidobacterium* and *Lactobacillus*, specifically *L. reuteri, L. acidophilus, L. casei, L. plantarum,* and *L. rhamnosus.*

Microbiotas can also be classified as probiotics because they convey health benefits on the host (the absence of a microbiota makes germfree mice more susceptible to infection than conventional mice). Furthermore, the use of antibiotics that damage the microbiota can lead to overgrowth by enteric pathogens such as *Salmonella enterica* serovar Typhimurium and, above all, *Clostridium difficile.*

Recent studies have begun to elucidate the molecular mechanisms behind the beneficial role of commensal bacteria and probiotics. There are two primary mechanisms involved. The first is direct competition for nutrients or an ecological niche. The second is indirect, in which the commensals and probiotics stimulate the immune system in an effect called *physiological inflammation*; in other words, commensal bacteria and probiotics stimulate a very low level of inflammation that interferes with colonization of the gut by pathogens if they attack.

Several studies have shown that nutrient and/or ecological niche competition exists between similar bacterial strains. For example, in mice treated with streptomycin to kill most of their commensal microbiota, colonization by certain strains of *Escherichia coli* (strains HS and Nissle) prevents colonization by enteropathogenic *E. coli* (EPEC). The ability of *E. coli* strains HS and Nissle to use multiple sugars as nutrients starves out the *E. coli* pathogens. Naturally, certain pathogens have found ways around this and succeed in colonization by using sugars that the commensal strains do not, or even by catabolizing sugars released by the microbiota itself.

E. coli strain Nissle is used as a human probiotic. It was isolated in 1917 from the feces of a soldier who remained healthy during an outbreak of shigellosis, a disease due to *Shigella,* a bacterium closely related to *E. coli.* This Nissle strain has become an important component in the preparation of probiotics used for treating diarrhea and intestinal inflammatory conditions such as Crohn's disease. In addition to its ability to use a variety of sugars, the Nissle strain has several iron collection systems that help it compete with pathogens. Similar to other commensal bacteria, it is also able to interact directly with pathogens by producing toxins and antimicrobial peptides called *microcins.*

Type VI secretion systems, which allow for the secretion of antibacterial toxins via a phage-type machinery, were once considered to be a virulence mechanism of pathogens against commensal bacteria. Curiously, however, it has now been found that the commensal bacteria use these systems themselves to attack invaders. The large proportion of *Bacteroidetes* in the intestinal microbiota has been attributed to the environmental advantage provided by their type VI secretion systems.

It must be stressed that, so far, most research in this area has been restricted to mice, and the application of these results to humans has yet to be determined. The use of "humanized" mice, mice endowed with human microbiotas, would be of great use for validating these early results.

In addition to the effects of competition for the same niche, probiotics and commensal bacteria also strengthen the barrier effect of the intestinal mucosa and increase innate immune responses and adaptive responses (as discussed in chapter 9).

Fecal transplants

As the value of the intestinal microbiota becomes more widely recognized, interest has turned to fecal transplants. In this technique, a sample of the intestinal microbiota from an individual in good health is introduced into the intestinal tract of a patient. Generally an effort is made to use the microbiota of a person in the same family as long as it is free of pathogens, whether viral, bacterial, or parasitic.

Fecal transplants are currently used primarily to treat *C. difficile* pseudomembranous colitis, a health care-related infection that can follow antibiotic treatments that damage the normal intestinal microbiota. Fecal transplants are also used to treat Crohn's disease and other intestinal inflammatory conditions.

The intestinal microbiota of insect vectors

In the framework of studies on the tripartite relationships between pathogens, their hosts, and insect vectors, progress on the human intestinal microbiota has sparked interest in the intestinal microbiotas of the tsetse fly and different types of mosquitoes.

The most abundant bacteria in insect microbiotas are proteobacteria of the *Enterobacteriaceae* family. Members of the genera *Enterobacter*, *Pantoea*, *Pseudomonas*, and *Serratia* are also found. With the exception of vertically transmitted bacteria, such as *Asaia* in mosquitoes and symbiotic bacteria in tsetse flies, how insects acquire bacteria is unknown.

Several studies have shown that the presence of symbiotic bacteria in mosquitoes can reduce their ability to act as vectors. Larger quantities of intestinal bacteria are associated with a lower rate of infection in *Anopheles* mosquitoes exposed to the malaria parasite *Plasmodium falciparum*. Supporting this hypothesis are studies that show that mosquitoes treated with antibiotics before they take a blood meal end up with a higher number of parasites in their gut.

However, not all bacteria have the same ability to reduce mosquitoes' infection rates. A study succeeded in identifying an *Enterobacter* species in Zambian mosquitoes that confers resistance to *P. falciparum* infection by interfering with the development of the parasite before it invades the insect's intestinal epithelium. This antiparasitic effect can be attributed to the release of oxygen radicals by the bacteria. These results suggest the possibility of manipulating the intestinal microbiota in mosquitoes to make them resistant to *P. falciparum*.

CRISPR/Cas9 and gene therapy

Gene therapy in humans is a technology that has, thus far, consisted of replacing defective genes with normal genes in cells and tissues where the genes are strongly expressed and play a key role. This technique uses viruses that are able to insert themselves into genomes, such as retroviruses, although the technique is not without risks.

The ease with which genomes can now be manipulated with the CRISPR/Cas9 system is tempting for adventurous spirits who dream of genetically manipulating germ cells in order to heal every type of cell in an individual. Two commentaries published in *Nature* and *Science* in March 2015, however, and signed by eminent scientists, including Nobel Prize winner David Baltimore, called on the scientific community not to use CRISPR to modify DNA in germ cells. The risk of a patchwork embryo, with some cells corrected and others not, is a concern, as well as the possibility of mutations occurring at locations other than those targeted. Finally, the commentaries argued that society must be informed as clearly as possible of the fact that while CRISPR is a powerful and enticing technique, there are currently too many risks, and that gene therapy in human somatic cells is a very different prospect from that in germ cells.

Synthetic biology

In the 1970s, the term "genetic engineering" referred to techniques in molecular and genetic biology that were used to isolate, clone, express, or overexpress genes in organisms different from those in which the gene originated. For example, genetic engineering was used to create *E. coli* strains that produced ovalbumin, human growth hormone, interferon, and insulin, as well as enzymes to be added to laundry detergents and proteins used to vaccinate children. One of the best examples of this last is undoubtedly the *Bordetella pertussis* toxin, produced by *E. coli*, that was combined with two other proteins, filamentous hemagglutinin and adhesin, as a vaccine against whooping cough (pertussis); this was the first "subunit" or acellular vaccine. Techniques using *Agrobacterium tumefaciens* and the Ti plasmid introduced genes into plants that provided resistance to herbicides or toxins to repel insect pests. These were the first transgenic plants, or genetically modified organisms (GMOs).

In the genetic engineering era, the host of choice was generally *E. coli*, as all other tools in molecular biology to that point had been developed in the same bacterium. CRISPR/Cas9 technology upset this norm, and thus synthetic biology has made its most notable advances in other organisms, including plants.

Synthetic biology is the "postgenomic" version of genetic engineering. A real revolution is taking place with the discovery, optimization, and production of bioactive molecules. As with genetic engineering, synthetic biology involves a modified microorganism—referred to as a *chassis*—to mass-produce compounds that are expensive, pharmacologically valuable, or difficult to synthesize chemically. Synthetic biology has profited from the flood of genomic and metagenomic data to extend the principles of genetic engineering to new and inventive commercial products.

Synthetic biology relies on the discovery of new biosynthetic pathways, whether they are active or silent in the microorganisms under study. It is also applied in the search for previously unknown and potentially useful microbial compounds, identified by recent new techniques such as mass spectrometry and its associated developments. Microorganisms produce many metabolites, small molecules that have a huge bioactive potential. Although many projects involving metabolites focus on the production of aromatic molecules for use in flavoring foods or perfuming cosmetics, the hope is that they will be used to produce medicines, particularly new antibiotics.

The most emblematic example of synthetic biology is the antimalarial drug artemisinin. Along with ivermectin, artemisinin was recognized by the Nobel committee and earned Chinese chemist Tu Youyou a Nobel Prize in 2015. Artemisinin can now be produced by heterologous organisms, particularly the baker's yeast *Saccharomyces cerevisiae*. Production in *E. coli* proved too complicated and inefficient; those studies took 10 years and involved too many steps in extracting the drug from its precursor, artemisinic acid.

Synthesis of valinomycin

Another emblematic achievement of synthetic biology is the synthesis of valino-mycin by *E. coli*. Produced by several *Streptomyces* species, valinomycin is a useful antibiotic and ionophore. Researchers have succeeded in introducing the coding locus for a nonribosomal peptide synthetase into *E. coli*, paving the way for the synthesis of similar modified valinomycins.

Synthetic biology has been used to activate silent biosynthetic pathways in *Streptomyces orinoci* to produce spectinabilin, a polyketide with both anti-malarial and antiviral activity. In order to achieve this, researchers had to remove all regulatory sequences that prevented the expression of the locus encoding the enzymes involved in the polyketide synthesis.

Another use for synthetic biology is in *de novo* genome assembly. Craig Venter and his colleagues succeeded in this goal in 2012 with the synthesis of a new organism. They synthesized *in vitro* (chemically) the genome for *Mycoplasma mycoides* and transplanted it into *Mycoplasma capricolum*. This process involved expressing DNA fragments in yeast and then transplanting them into the bacteria. At the conclusion of the experiment, *M. capricolum* bacteria that now contained only the *M. mycoides* genome replicated using the genome of *M. mycoides* . Venter called these cells "synthetics," a slight approximation since their cytoplasm did derive from the original recipient cell. Venter then progressed to the next step, removing from this artificial genome all sequences that were nonessential for survival. With 473 genes, the new synthetic bacterium, which replicates every 3 hours, is the simplest known form of life.

Among the many advances based on synthetic biology, one can cite the creation of "semisynthetic" organisms, capable of using a new and different base pair with two nonnatural bases (d5SICS and dNaM). We know that in genomes, DNA is a double strand formed by two fundamental base pairs: A plus T and G plus C (see Fig. 5). In order for the organism to use this novel base pair, it was necessary to create a transporter that could introduce the required nucleotides into the genome; again, this was accomplished in *E. coli*. It was shown that the replication machinery of the new organism used these new compounds and did not introduce mutations. Furthermore, the repair enzymes that normally excise anomalous bases in DNA did not attack the new base pair. This newly generated organism, now able to utilize *three* different base pairs, is the first of its kind. It should be able to serve as a host for synthesizing important pharmaceutical products as well as other industrial purposes, with the extra security of the fact that it cannot possibly replicate in nature.

CHAPTER 19

Bacteria as Environmental Tools

Bacillus thuringiensis as a biopesticide

Bacillus thuringiensis was discovered in Japan in silkworms in 1902 and was later isolated from flour moths in Thüringen, Germany, in 1911. *B. thuringiensis* is an entomopathogenic bacterium that was rapidly adopted worldwide as a biopesticide thanks to its toxicity to the larvae of insects, including butterflies, beetles, flies, hymenoptera, homoptera, and lice, as well as certain invertebrates. The *B. thuringiensis* toxin, often referred to as *Bt toxin*, is made up of the proteins Cry and Cyt, two delta-endotoxins that are generated during sporulation. More than 600 *cry* genes have so far been identified. Beyond its unique ability to produce Cry toxins, *B. thuringiensis* is very similar to *Bacillus anthracis*, the bacterial agent of anthrax, and to *Bacillus cereus*, an environmental bacterium that can cause foodborne disease. *B. thuringiensis* produces other factors that contribute to its usefulness as an insecticide, including chitinases, proteases, and other toxins. The specificity of Bt's action is related to the specificity of its toxins.

The toxin crystals produced by *B. thuringiensis* during sporulation are *protoxins*; in other words, these protoxins, when ingested by the insect, mature in the alkaline pH of the insect gut into active polypeptide toxins that attach to specific receptors on the host's intestinal epithelial cells. Their effect is rapid. The toxins cause large intestinal lesions and paralysis of the digestive tube, and the insect dies within 48 hours.

The first successful uses of Bt as a biopesticide date from the 1960s in the United States and the 1970s in France. It is the most widely used biopesticide. It is easily cultivated in fermentors, and the final product is stable, highly selective, and competitively priced. Bt has no known effects on useful fauna such as pollinators (bees) or vertebrates. Resistance to Bt toxin first appeared in 1990 in a strain of *Plutella xylostella* (cabbage or diamondback moth) isolated in Hawaii. In most cases, resistance is due to mutations in the genes that encode the toxin receptors on the insect's intestinal cells.

An important discovery that has reshaped perspectives on Bt use was the creation of plants genetically modified to produce the Bt toxin. This was first demonstrated in a tobacco plant modified to express a toxin active against the tobacco hawk moth *Manduca sexta*. Following this work, many other plants (tomatoes, corn, cotton) have been manipulated to give them insecticide-like properties. There is still significant resistance to the use of genetically modified plants or GMOs.

Much current research is focusing on *B. thuringiensis* strains that have insecticide-like properties against insects such as the *Aedes aegypti* mosquito, a member of the *Culicidae* family that preys on animals and humans, and against *Ceratitis capitata* or medfly, a fly of the *Tephritidae* family that feeds on certain fruit trees native to the Mediterranean.

Bacillus subtilis to protect plant roots

Certain strains of *Bacillus subtilis*, the Gram-positive model bacterium (*Escherichia coli* is the Gram-negative model bacterium), secrete a surfactin, an antimicrobial lipopeptide that contributes to the formation of biofilms. These biofilms grow on the surface of plant roots and protect them against attacks from pathogens, as shown in studies of *Arabidopsis* (cress) infection with *Pseudomonas syringae*. The *B. subtilis* GBO3 strain has been commercialized for use on flowers, cotton, vegetables, and soybeans. This bacterium is a sporeformer; plant seeds are coated with the spores, and when the seeds germinate, the spores activate and develop their root-protective biofilms.

Wolbachia and biocontrol of mosquito-borne infectious diseases

Once established in certain mosquitoes, *Wolbachia* bacteria inhibit the transmission of several viruses, including those responsible for dengue, chikungunya, and yellow and West Nile fevers, as well as the malaria parasite *Plasmodium falciparum*. Furthermore, *Wolbachia* causes cytoplasmic incompatibility, a form of sterility in insects (see Fig. 15). This bacterium is the basis for several strategies to eliminate populations of mosquitoes or other insects that are considered highly dangerous. The two main approaches currently used include replacing wild mosquitoes with those carrying *Wolbachia* (making them resistant to pathogens) and eliminating populations of pathogen-carrying mosquitoes.

Replacing wild mosquitoes with *Wolbachia* carriers

In this first strategy, female mosquitoes infected with *Wolbachia* are released into the environment. Due to the *Wolbachia*, they are resistant to many pathogens and therefore play host to far fewer other pathogens than do wild mosquitoes. When infected female mosquitoes mate with uninfected males, their offspring are both viable and infected, increasing the number of *Wolbachia* carriers in the environment.

Only females transmit *Wolbachia*. If an uninfected female mates with an infected male, they do not produce offspring, while infected females can produce viable, infected offspring with both infected and uninfected males (see Fig. 15). Releasing infected females thus results in an increase of infected mosquitoes of both sexes. Introducing enough infected females at once or repeatedly over a period of time should result in eventual replacement of an uninfected population by a *Wolbachia*-infected population that does not transmit pathogens. This was proven effective in fighting dengue virus by introducing *Wolbachia* into the mosquito A. *aegypti*.

Eliminating carrier mosquitoes

The second strategy that uses *Wolbachia* bacteria for biocontrol of mosquito-borne infectious diseases involves eliminating mosquitoes or reducing their populations by targeting the male mosquitoes. An older version of this strategy involved releasing males sterilized by irradiation into the environment to reduce the population. This was successful in completely eradicating the insect vector for filariasis, *Culex pipiens*. A more recent, *Wolbachia*-dependent version of this strategy involves the release of males infected with *Wolbachia*. When they mate with noninfected females, the offspring fail to hatch. Introducing infected male mosquitoes can thus lead to a decrease in the population of mosquitoes capable of carrying pathogens.

CHAPTER 20

Conclusion

In this book, I have intended to show that the whole field of microbiology, and in particular bacteriology, is in a vibrant and dynamic phase of growth. Many new concepts have appeared, and many others are about to emerge. This revolution will influence our daily lives, our diet, and our health care, as well as multiple fields of research in biology, medicine, agriculture, and industry and, it may be hoped, the protection of our environment.

I have discussed in great detail the major problem of bacterial resistance to antibiotics and tried to show that alternative methods of disease eradication are possible. A new antibiotic, teixobactin, was discovered in 2016 thanks to novel techniques, and it is more than likely that others will soon be discovered. When resistance to the antimalarial nivaquine appeared, an alternative drug, artemisinin, was first extracted from a plant and then produced by synthetic biology. The success of artemisinin as a replacement for nivaquine, among many other discoveries, raises hopes that new antibacterial therapies will come to light.

I have emphasized the major role of all the assemblies of bacteria and other microorganisms that are called microbiotas, which actively contribute to our health and lifestyle throughout our lives. These microbiotas, the intestinal microbiota in particular, stimulate and control a variety of functions, especially immune system defenses against pathogens and the production of hormones such as serotonin. Research shows a clear correlation between having a highly diversified intestinal microbiota and maintaining a healthy, balanced mental and physical state of well-being. The intestine controls what happens in the body, like a "second brain." Will we be able to manipulate patients' intestinal microbiotas to help cure illnesses? This already seems to be the case for inflammatory conditions such as Crohn's disease. It is only a little more difficult to imagine that we might be able to do the same for mental conditions such as depression. Why not? In any case, because research suggests that our diets contribute to the diversity of our intestinal microbiota, and given the value of a diverse microbiota to our health, here is some advice. Never pass up the opportunity to have a good meal, but don't forget the basic rules of a well-balanced diet!

Fecal transplantation is a new technology that could benefit individuals with dysbiosis or imbalances in their intestinal microbiota. It is already used successfully to treat diarrhea caused by *Clostridium difficile*, the bacteria that can cause gastrointestinal superinfection following antibiotic treatment. Fecal autotransplantation could even be planned if antibiotic treatment is anticipated far enough in advance to allow time for the collection of an autologous fecal sample. That said, fecal transplantation has already been the subject of much debate by regulatory committees over whether it should be considered tissue grafting or medical intervention, and different agencies have come to different conclusions.

As I have tried to illustrate throughout this book, a wide range of key discoveries with broad-reaching consequences have been made by researchers investigating fundamental mechanisms of bacterial resistance to phages and/or regulatory mechanisms. As demonstrated by the recent studies on regulatory RNA and CRISPR/Cas9 systems, unrestricted basic research should be not only maintained but encouraged. Nobody could have predicted that the results of such research would lead to a revolutionary technology of genome modification that could lead to gene therapy.

I have mentioned insects often, both as vectors for pathogens in humans (specifically mosquitoes), plants (green fly, aphids), and animals and also as important agricultural concerns, from pollinators (bees) to pests. I have discussed their microbiotas and the role of their intestinal microbiota in controlling their ability to act as carriers of pathogens. Will it be useful to try to modify the intestinal microbiota of insects? Studies of endosymbionts in insects—particularly *Wolbachia*, which is transmitted from one generation to the next in oocytes and affects mosquitoes' ability to carry pathogens—have given rise to the idea of releasing infected mosquitoes into the environment, an effort already proven successful in reducing and eradicating pathogens.

With the advent of global climate change, it is safe to bet that insect populations will change. This has already been observed with tiger mosquitoes. Certain insects will become more and more prevalent in areas they currently do not occupy. Problems, possibly critical, will arise that are under control or nonexistent at this time, such as infections with phytoplasmas, which are crop-destroying bacteria that are spread by leafhoppers. Fortunately, our knowledge of bacteria continues to grow, allowing us to deal with many threats and even anticipate new ones while maintaining the biodiversity of the environment that we are coming to know better and better.

Major Figures in Microbiology

Antonie van Leeuwenhoek (1632–1723) used microscopes capable of 300-fold magnification to observe many microorganisms and protozoa. He was also the first person to observe living sperm cells. He was a vigorous opponent of the theory of spontaneous generation.

Louis Pasteur (1822–1895), chemist, demonstrated the role of yeast in fermentation, quashed the theory of spontaneous generation, and introduced the concept of pasteurization, or heating, of beer, wine, or food in order to eliminate bacterial contamination. He identified the pathogen responsible for pébrine, or *pepper disease*, in silkworms, then focused his attention on vaccinating chickens against cholera and humans against the rabies virus. At the end of his life, he called for donations from the international public and founded the Pasteur Institute, created by decree on June 4, 1887.

Robert Koch (1843–1910) (Nobel Prize 1905), physician, considered as the founder of modern microbiology, demonstrated that anthrax is caused by spores of *Bacillus anthracis*; discovered the bacillus responsible for tuberculosis, often referred to as Koch's bacillus; and discovered *Vibrio cholerae*, the bacterium responsible for cholera. He established the scientific postulates that bear his name, used to this day to identify the causative agent of a given disorder. He finished his career with an interest in tropical diseases.

Hans Christian Gram (1853–1938) developed in 1884 the coloration technique that bears his name (Gram stain) and is still the most widely used method for classifying bacteria into two groups: Gram-positive bacteria and Gram-negative bacteria.

Alexander Fleming (1881–1955) (Nobel Prize 1945) discovered around 1929 the antibacterial properties of penicillin, produced from a *Penicillium* species, and ushered humanity into the age of antibiotics.

Albert Schatz (1920–2005) and Selman Waksman (1888–1973) (Waksman, 1952 Nobel Prize) discovered in 1943 another antibiotic, streptomycin, which started to be used in humans against tuberculosis in 1949. Unfortunately, the first resistance to antibiotics began to appear not long afterward, in 1946 for penicillin and in 1959 for streptomycin.

François Jacob (1920–2013), Jacques Monod (1910–1976), and André Lwoff (1902–1994) (Nobel Prize 1965) proposed in 1960 the concept of the "operon" to explain the coordinated control of bacterial genes by proteins they termed *repressors* that attach to sites called *operators*.

Carl Woese (1928–2012) studied ribosomal RNA and in 1977 thereby discovered archaea, a third category of life form genetically distinct from bacteria and eukaryotes.

Stanley Falkow (1934–) was one of the first to combine tools used in genetics with cellular biology, which led him to identify virulence factors in pathogenic bacteria.

Kary Mullis (1944–) (Nobel Prize 1993) invented PCR (polymerase chain reaction) technology with an enzyme found in the thermophilic bacterium *Thermus aquaticus*. PCR has become a fundamental tool in molecular biology.

Craig Venter (1946–) and his colleagues at The Institute for Genomic Research determined the first complete sequence of a bacterial genome (that of *Haemophilus influenzae*), ushering microbiology into the era of genomics.

Jeffrey I. Gordon (1947–) pioneered research on intestinal microbiota and its role in conditions such as obesity.

Jennifer Doudna (1964–), Emmanuelle Charpentier (1968–), Philippe Horvath (1970–), and many others over the past several years have participated in the development of the CRISPR/Cas9 genome editing technology.

GLOSSARY

Antibiotic: A compound that kills or inhibits growth of a bacterium.

Archaea: The third domain of life; the other two are the *Bacteria* and *Eukarya*.

ATP: Adenosine triphosphate. This compound is made of different parts connected by bonds that liberate energy when they are disrupted. In humans, ATP is involved in muscular contraction.

Bacilli: Bacteria that have an elongated form.

Bacteriocin: A toxin made by a bacterium that kills other bacteria.

Bacteriophage: A virus that targets bacteria.

Bacterium: A single-celled organism, the smallest living organism.

Biofilm: A community of microbes living on a surface and forming a discrete structure. Biofilms comprise many cells from one or many species within an extracellular matrix.

Chromosome: The DNA molecule that contains the genes of an organism. In bacteria, the chromosome is generally circular.

Cocci: Bacteria that have a nearly spherical shape.

Competence: A special physiological state that can occur in some bacteria, at a specific growth phase, or in specific growth conditions, and allows bacteria to take up DNA from the environment.

Conjugation: A phenomenon whereby a bacterium can transfer part of its DNA or its plasmid to another bacterium via a nanotube through which the DNA passes.

CRISPR: Acronym for clustered regularly interspaced short palindromic repeats. A CRISPR region is a region of the bacterial chromosome that contains repeats of about 50 nucleotides and spacers made of phage or plasmid DNA. This region can integrate a piece of phage DNA upon a first infection with a given virus and can later interfere with an infection with the same virus. The phenomenon is compared to a "memory" of the first infection.

DNA: Deoxyribonucleic acid. Composed of two long strands made of nucleotide subunits, DNA is the basic component of the chromosomes in all organisms, including bacteria and higher organisms. Hence it is the heritable material.

DNA polymerase: An enzyme that produces DNA molecules by copying the DNA during the process of replication.

Gene transfer: Transfer of a gene, i.e., a piece of DNA, from one bacterium to another by a specific mechanism: via plasmid conjugation, transformation, or infection by a bacteriophage.

Gene: A region of DNA (i.e., of the chromosome) that encodes the information for a single protein.

Genome: The sum of all genes of a particular organism. Generally, in bacteria, all genes are located on one single circular DNA molecule called a chromosome.

High-throughput DNA sequencing: Sequencing of all DNA molecules contained in a particular extract.

LPS: Liposaccharide. Located on the external face of the outer membrane of Gram-negative bacteria, this component stimulates the innate immune system with high efficiency. It is therefore called an endotoxin.

Microbe: A generic name given to any microorganism: a bacterium, a protozoan, a yeast. This word is also sometimes used to refer to an infectious agent of unknown origin.

Microbial community: A community of microbial species living together in a given environment.

Microbial communication: Interactions between bacteria or microorganisms that are mediated by chemical compounds and result in the production of different compounds or in change in behavior.

Microbiome: This word describes either the ensemble of genes present in a given environment or both the genes and the molecules that surround them, i.e., the microbial ecosytem.

Microbiota: The ensemble of all microorganisms present in a given location, e.g., the intestinal microbiota or the skin microbiota.

Multicellular organism: An organism made of a high number of cells of different kinds.

Nosocomial infection: Also referred to as "health care-related infection," an infection acquired at the hospital or due to other medical intervention.

Nucleotide: The name of the subunit of the nucleic acids DNA or RNA.

Oocyte: A female cell in an ovary that may mature into a germ cell and allows reproduction in higher organisms.

Operon: A small group of genes that are coexpressed, i.e., transcribed together and regulated together.

Pathogenicity island: A group of genes clustered in a single locus on the chromosome and involved in the pathogenicity of a bacterium.

Peptidoglycan: A component of the bacterial cell wall made of sugars and amino acids that confers to the bacterium rigidity and protection against pathogens or external attacks.

Plasmid: Small circular chromosome made of two strands of DNA.

Quorum sensing: A phenomenon mediated by chemical communication between bacteria. It allows them to behave as a multicellular organism in a coordinated manner by responding together to an accumulation of small molecules. Quorum sensing allows microbes to indirectly sense and evaluate the number of other microbes present in the environment.

Replication: The process by which one DNA strand is copied by the DNA polymerase.

Ribosome: The complex of proteins and RNA molecules that synthesizes proteins in all living cells.

RNA: Ribonucleic acid. This compound is a single strand made of nucleotide subunits. RNA is formed upon transcription of DNA.

RNA polymerase: An enzyme that produces RNA molecules by copying the DNA nearly exactly during the process called transcription.

Spores: Structures that form following bacterial starvation or harsh growth conditions. They allow bacteria to survive under extreme environmental conditions, then germinate when their environment is more advantageous.

Symbiont: One organism living in symbiosis with another with mutual benefit.

Transcription: The process by which the bacterial machinery, in particular the enzyme RNA polymerase, reads the DNA and produces a nearly exact copy, the RNA.

Translation: The process by which the bacterial ribosome reads the RNA and produces proteins.

Transformation: The phenomenon that allows bacteria to take up DNA present in the environment. Transformation can only occur in competent bacteria. A certain number of bacteria are naturally competent. Competence can also be induced artificially in the laboratory in order to introduce DNA.

Vacuole: A vesicle-like organelle within a cell, formed, for example, when a bacterium enters a mammalian cell. The bacterium is then trapped in a membrane-enclosed compartment, the vacuole.

Virus: Infectious agent made of DNA or RNA and proteins. Viruses infect organisms and depend on their hosts in order to replicate.

BIBLIOGRAPHY

Preface

Radoshevich L, Bierne H, Ribet D, Cossart P. 2012. The new microbiology: a conference at the Institut de France. *C R Biol* **335**:514–518.

PART I:
New Concepts in Microbiology

Chapter 1
Bacteria: Many Friends, Few Enemies

Woese CR, Fox GE. 1977. Phylogenetic structure of the prokaryotic domain: the primary kingdoms. *Proc Natl Acad Sci USA* **74**:5088–5090.

Ciccarelli FD, Doerks T, von Mering C, Creevey CJ, Snel B, Bork P. 2006. Toward automatic reconstruction of a highly resolved tree of life. *Science* **311**: 1283–1287.

Medini D, Serruto D, Parkhill J, Relman DA, Donati C, Moxon R, Falkow S, Rappuoli R. 2008. Microbiology in the post genomic era. *Nat Rev Microbiol* **6**:419–430.

Chapter 2
Bacteria: Highly Organized Unicellular Organisms

Jensen RB, Wang SC, Shapiro L. 2002. Dynamic localization of proteins and DNA during a bacterial cell cycle. *Nat Rev Mol Cell Biol* **3**:167–176.

Gitai Z. 2005. The new bacterial cell biology: moving parts and cellular architecture. *Cell* **120**:577–586.

Cabeen MT, Jacobs-Wagner C. 2007. Skin and bones: the bacterial cytoskeleton, cell wall, and cell morphogenesis. *J Cell Biol* **179**:381–387.

Cabeen MT, Jacobs-Wagner C. 2010. The bacterial cytoskeleton. *Annu Rev Genet* **44**:365–392.

159

Toro E, Shapiro L. 2010. Bacterial chromosome organization and segregation. *Cold Spring Harb Perspect Biol* **2:**a000349.

Campos M, Jacobs-Wagner C. 2013. Cellular organization of the transfer of genetic information. *Curr Opin Microbiol* **16:**171–176.

Ozyamak E, Kollman JM, Komeili A. 2013. Bacterial actins and their diversity. *Biochemistry* **52:**6928–6939.

Laoux G, Jacobs-Wagner C. 2014. How do bacteria localize proteins to the cell pole? *J Cell Sci* **127:**11–19.

Chapter 3
The RNA Revolution

Jacob F, Monod J. 1961. Genetic regulatory mechanisms in the synthesis of proteins. *J Mol Biol* **3:**318–356.

Roth A, Breaker RR. 2009. The structural and functional diversity of metabolite-binding riboswitches. *Annu Rev Biochem* **78:**305–309.

Gottesman S, Storz G. 2011. Bacterial small regulators: versatile roles and rapidly evolving variations. *Cold Spring Harb Perspect Biol* **3:**pii:a003798. doi:10.1101/cshperspect.a003798.

Storz G, Vogel J, Wasserman KM. 2011. Regulation by small RNAs in bacteria: expanding frontiers. *Mol Cell* **43:**880–891.

Breaker RR. 2012. Riboswitches and the RNA world. *Cold Spring Harb Perspect Biol* **4:**pii:a003566. doi:1101/cshperpect.a003566.

Calderi I, Chao Y, Romby P, Vogel J. 2013. RNA-mediated regulation in pathogenic bacteria. *Cold Spring Harb Perspect Biol* **3:**a010298. doi:10.1101/cshperpect.a010298.

Sesto N, Wurtzel O, Archambaud C, Sorek R, Cossart P. 2013. The excludon: a new concept in bacterial anti-sense RNA mediated gene regulation. *Nat Rev Microbiol* **11:**75–82.

Mellin JR, Tiensuu T, Becavin C, Gouin E, Johansson J, Cossart P. 2013. A riboswitch-regulated anti-sense RNA in *Listeria* monocytogenes. *Proc Natl Acad Sci USA* **110:**13132–13137.

Chapter 4
From the CRISPR Defense System to the CRISPR/Cas9 Method
for Modifying Genomes

Barrangou R, Fremaux C, Deveau H, Richards M, Boyaval P, Moineau S, Romero DA, Horvath P. 2007. CRISPR provides acquired resistance against viruses in prokaryotes. *Science* **315:**1709–1712.

Deltcheva E, Chylinski K, Sharma S, Gonzales K, Chao Y, Pirzada ZA, Eckert MR, Vogel J, Charpentier E. 2011. CRISPR RNA maturation by *trans*-encoded small RNA and host factor RNAse III. *Nature* **471:**602–607.

Jinek M, Chylinski K, Fonfara I, Hauer M, Doudna JA, Charpentier E. 2012. A programmable dual-RNA-guided DNA endonuclease in adaptive bacterial immunity. *Science* 337:816–821.

Jiang W, Bikard D, Cox D, Zhang F, Maraffini LA. 2013. RNA-guided editing of bacterial genomes using CRISPR-Cas systems. *Nature Biotech* 31:233–239.

Dupuis ME, Villion M, Magadan AH, Moineau S. 2013. CRISPR-Cas and restriction-modification systems are compatible and increase phage resistance. *Nat Commun* 4:2087.

Hsu P, Lander E, Zhang F. 2014. Development and applications of CRISPR-Cas9 for genome editing. *Cell* 157:1262–1278.

Selle K, Barrangou R. 2015. Harnessing CRISPR-Cas systems for bacterial genome editing. *Trends Microbiol* 23:225–232.

Kiani S, Chavez A, Tuttle M, Hall RN, Chari R, Ter-Ovanesyan D, Qian J, Pruitt BW, Beal J, Vora S, Buchthal J, Kowal EJ, Ebrahimkhani MR, Collins JJ, Weiss R, Church G. 2015. Cas9 gRNA engineering for genome editing, activation and repression. *Nat Methods* 11:1051–1054.

Chapter 5
Antibiotic Resistance

Sockett E, Lambert C. 2004. *Bdellovibrio* as therapeutic agents: a predatory renaissance. *Nat Rev Microbiol* 2:669–674.

Dublanchet A, Fruciano E. 2008. Brève histoire de la phagothérapie. A short history of phage therapy. *Med Maladies Infect* 38:415–420.

Debarbieux L, Dublanchet A, Patay O. 2008. Infection bactérienne: quelle place pour la phagothérapie. *Med Maladies Infect* 38:407–409.

Makarov V, et al. 2009. Benzothiazinones kill *Mycobacterium tuberculosis* by blocking arabinan synthesis. *Science* 8:801–804.

Cotter P, Ross RP, Hill C. 2013. Bacteriocins—a viable alternative to antibiotics. *Nat Rev Microbiol* 11:95–105.

World Health Organization. 2014. WHO's first global report on antibiotic resistance reveals serious, worldwide threat to public health. April 2014. http://www.who.int/mediacentre/news/releases/2014/amr-report/en/ [Premier rapport de l'OMS sur la résistance aux antibiotiques: une menace grave d'ampleur mondiale, avril 2014, http://www.who.int/mediacentre/ news/releases/2014/amr-report/fr/].

Institut Pasteur. 2014. Antibiotiques: quand les bactéries font de la résistance. (Dossier.) *Lettre de l'Institut Pasteur* 85. https://www.pasteur.fr/sites/default/files/rubrique_nous_soutenir/lip/lip85-resistance_aux_antibiotiques-institut-pasteur.pdf (2014).

Lambert C, Sockett RE. 2013. Nucleases in *Bdellovibrio bacteriovorus* contribute towards efficient self-biofilm formation and eradication of preformed prey biofilms. *FEMS Microbiol Lett* 340:109–116.

Allen H, Trachsel J, Looft T, Casey T. 2014. Finding alternatives to antibiotics. *Ann NY Acad Sci* **1323:**91–100.

Baker S. 2015. A return to the pre-antimicrobial era? The effects of antimicrobial resitance will be felt most acutely in lower income countries. *Science* **347:**1064.

Ling L, Schenider T, Peoples A, Spoering A, Engels I, Conlon BP, Mueller A, Schäberle TF, Hughes DE, Epstein S, Jones M, Lazarides L, Steadman V, Cohen DR, Felix C, Fetterman KA, Millet W, Nitti AG, Zullo AM, Chen C, Lewis K. 2015. A new antibiotic kills pathogens without detectable resistance. *Nature* **517:**455–459.

PART II:
Sociomicrobiology: The Social Lives of Bacteria

Chapter 6
Biofilms: When Bacteria Gather Together

Davies DG, Parsek MR, Pearson JP, Iglewski BH, Costerton JW, Greenberg EP. 1998. Involvement of cell-to-cell signals in the development of a bacterial biofilm. *Science* **280:**295–298.

O'Toole G, Kaplan HB, Kolter R. 2000. Biofilm formation as microbial development. *Annu Rev Microbiol* **18:**49–79.

Stanley NR, Lazazzera BA. 2004. Environmental signals and regulatory pathways that influence biofilm formation. *Mol Microbiol* **52:**917–924.

Kolter R, Greenberg EP. 2006. Microbial sciences: the superficial life of microbes. *Nature* **441:**300–302.

Römling U, Galperin MY, Gomlesky M. 2013. Cyclic di-GMP: the first 25 years of a universal bacterial second messenger. *Microbiol Mol Biol Rev* **77:**1–52.

Chapter 7
How Bacteria Communicate: Chemical Language and Quorum Sensing

Bassler BL, Losick R. 2006. Bacterially speaking. *Cell* **125:**237–246.

Duan F, March JC. 2008. Interrupting *Vibrio cholerae* infection of human epithelial cells with engineered commensal bacterial signaling. *Biotechnol Bioeng* **101:** 128–134.

Duan F, March JC. 2010. Engineered bacterial communication prevents *Vibrio cholerae* virulence in an infant mouse model. *Proc Natl Acad Sci USA* **107:** 11260–11264.

Schuster M, Sexton DJ, Diggle SP, Greenberg EP. 2013. Acyl-homoserine lactone quorum sensing: from evolution to application. *Annu Rev Microbiol* **67:**43–63.

Chapter 8
When Bacteria Kill Each Other

Schwarz S, West TE, Boyer F, Chiang WC, Carl MA, Hood RD, Rohmer L, Tolker-Nielsen T, Skerret S, Mougous J. 2010. *Burkholderia* type VI secretion systems have distinct roles in eukaryotic and bacterial cell interactions. *PLoS Pathog* 6:e10011068.

Hibbing ME, Fuqua C, Parsek M, Peterson SB. 2010. Bacterial competition: surviving and thriving in the microbiological jungle. *Nat Rev Microbiol* 8:15–25.

Hayes CS, Aoki SK, Low DA. 2010. Bacterial contact-dependent delivery systems. *Annu Rev Genet* 44:71–90.

Aoki S, Poole SJ, Hayes C, Low D. 2011. Toxin on a stick. Modular CDI toxin delivery systems play roles in bacterial competition. *Virulence* 2:356–359.

Russell AB, Hood R, Bui NK, LeRoux M, Vollmer W, Mougous J. 2011. Type VI secretion delivers bacteriolytic effectors to target cells. *Nature* 475:343–347.

Basler M, Ho BT, Mekalanos J. 2013. Tit for tat: type VI secretion system counterattack during bacterial cell-cell interactions. *Cell* 152:884–894.

Ho BT, Dong TG, Mekalanos JJ. 2014. A view to a kill: the bacterial type VI secretion system. *Cell Host Microbe* 15:9–21.

Etayash H, Azmi S, Dangeti R, Kaur K. 2015. Peptide bacteriocins. *Curr Top Med Chem* 16:220–241.

Chapter 9
Human-Animal Symbioses: The Microbiotas

McFall-Ngai M, Montgomery MK. 1990. The anatomy and morphology of the adult bacterial light organ of *Euprymna scolopes* (Cephalopoda: Sepiolidae). *Biol Bull* 179:332–339.

McFall-Ngai M, Heath-Heckman EA, Gillette AA, Peyer SM, Harvie EA. 2012. The secret languages of coevolved symbioses: insight from the *Euprymna scolopes–Vibrio fischeri* symbiosis. *Semin Immunol* 24:3–8.

McFall-Ngai M, Hadfield MG, Bosch TC, Carey HV, Domazet-Loso TX. 2013. Animals in a bacterial world, a new imperative for the life sciences. *Proc Natl Acad Sci USA* 110:3229–3236.

David LA, Maurice CF, Carmody RN, Gootenberg DB, Button JE, Wolfe BE, Ling AV, Devlin AS, Varma Y, Fischbach MA, Biddinger MA, Dutton EJ, Turnbaugh PJ. 2014. Diet rapidly and reproducibly alters the human microbiome. *Nature* 505:559–563.

Yurist-Doutsch S, Arrieta MC, Vogt SL, Finlay BB. 2014. Gastrointestinal microbiota-mediated control of enteric pathogens. *Annu Rev Genet* 48:361–382.

Brune A. 2014. Symbiotic digestion of lignocellulose in termite guts. *Nature Rev Microbiol* 12:168–180.

Belkaid Y, Segre J. 2014. Dialogue between skin microbiota and immunity. *Science* **346:**954–959.

Knights D, Ward T, McKinlay CE, Miller H, Gonzalez A, McDonald D, Knight R. 2014. Rethinking "enterotypes." *Cell Host Microbe* **16:**433–437.

Vogt SL, Pena-Diaz J, Finlay BB. 2015. Chemical communication in the gut: effects of microbiota-generated metabolites on gastrointestinal bacterial pathogens. *Anaerobe* **34:**106–115.

Derrien M, Van Hylckama Vlieg JET. 2015. Fate, activity and impact of ingested bacteria within the human gut. *Trends Microbiol* **23:**354–366.

Thompson JA, Oliveira RA, Djukovic A, Ubeda C, Xavier KB. 2015. Manipulation of the quorum sensing signal AI-2 affects the antibiotic-treated gut microbiota. *Cell Rep* **10:**1861–1871.

Asher G, Sassone-Corsi P. 2015. Time for food: the intimate interplay between nutrition, metabolism and the circadian clock. *Cell* **161:**84–92.

Yano J, Yu K, Donalsdson GP, Shastri GG, Phoebe A, Ma L, Nagler CR, Ismagilov RF, Mazmanian SK, Hsiao E. 2015. Indigenous bacteria from the gut microbiota regulate host serotonin biosynthesis. *Cell* **161:**264–276.

Schnupf P, Gaboriau-Routhiau V, Gros M, Friedman R, Moya-Nilges M, Nigro G, Cerf-Bensussan N, Sansonetti PJ. 2015. Growth and host interaction of mouse segmented filamentous bacteria in vitro. *Nature* **520:**99–103.

Sender R, Fuchs S, Milo R. 2016. Are we really outnumbered? Revisiting the ratio of bacterial to host cells in humans. *Cell* **164:**337–340.

Chapter 10
Bacterium-Plant Symbioses: Microbiotas of Plants

Jones KM, Kobayashi H, Davies BW, Taga ME, Walker GC. 2007. How rhizobial symbionts invade plants: the *Sinorhizobium-Medicago* model. *Nat Rev Microbiol* **5:**619–633.

Kondorosi E, Mergaert P, Kereszt A. 2013. A paradigm for endosymbiotic life: cell differentiation of *Rhizobium* bacteria provoked by host plants. *Annu Rev Microbiol* **67:**611–628.

Bulgarelli D, Schlaeppi K, Spaepen S, Ver Loren van Themaat E, Schulze-Lefert P. 2013. Structure and functions of the bacterial microbiota of plants. *Annu Rev Plant* **64:**807–838.

Chapter 11
Endosymbiotic Relationships

Lai CY, Baumann L, Baumann P. 1994. Amplification of TrpEG: adaptation of *Buchnera aphidicola* to an endosymbiotic association with aphids. *Proc Natl Acad Sci USA* **91:**3819–3823.

Douglas AE. 1998. Nutritional interactions in insect-microbial symbiosies: aphids and their symbiotic *Buchnera*. *Annu Rev Entomol* **43**:17–37.

Moran NA, Baumann P. 2000. Bacterial endosymbionts in animals. *Curr Opin Microbiol* **2**:270–275.

Gil R, Sabater-Munoz B, Latorre A, Silva FJ, Moya A. 2002. Extreme genome reduction in *Buchnera* spp.: toward the minimal genome needed for symbiotic life. *Proc Natl Acad Sci USA* **99**:4454–4458.

Sassera D, Beninati T, Bandi C, Bouman EAP, Sacchi L, Fabbi M, Lo N. 2006. *Candidatus Midichloria mitochondrii*, an endosymbiont of the tick *Ixodes ricinus* with a unique intramitochondrial lifestyle. *Internat J Systemat Evol Microbiol* **56**:2535–2540.

Moya A, Pereto J, Gil R, Latorr A. 2008. Learning how to live together: genomic insights into prokaryote-animal symbioses. *Nature Rev Genet* **8**:218–229.

Engelstadter J, Hurst GDD. 2009. The ecology and evolution of microbes that manipulate host reproduction. *Annu Rev Ecol Evol Syst* **40**:127–149.

Shigenobu S, Wilson ACC. 2011. Genomic revelations of a mutualism: the pea aphid and its obligate symbiont. *Cell Mol Life Sci* **68**:1297–1309.

Bouchery T, Lefoulon E, Karadjian G, Nieguitsila A, Martin C. 2012. The symbiotic role of *Wolbachia* in onchocercidae and its impact on filariasis. *Clin Microbiol Infect* **19**:131–140.

Scott AL, Ghedin E, Nutman TB, McReynolds LA, Poole CB, Slatko BE, Foster JM. 2012. Filarial and *Wolbachia* genomics. *Parasite Immunol* **34**:121–129.

Schulz F, Horn M. 2015. Intranuclear bacteria: inside the cellular control center of eukaryotes. *Trends Cell Biol* **25**:339–346.

PART III:
The Biology of Infections

Chapter 12
Pathogenic Bacteria, Major Scourges, and New Diseases

Shea JE, Hensel M, Gleeson C, Holden DW. 1996. Identification of a virulence locus encoding a second type III secretion system in *Salmonella typhimurium*. *Proc Natl Acad Sci USA* **93**:2593–2597.

Sansonetti PJ. 2006. The bacterial weaponry: lessons from *Shigella*. *Ann NY Acad Sci* **1072**:307–312.

Sussman M. (ed.). 2014. *Molecular Medical Microbiology*, 2nd ed. Academic Press, New York, NY.

Cornelis GR, Wolf-Watz H. 1997. The *Yersinia* Yop virulon: a bacterial system for subverting eukaryotic cells. *Mol Microbiol* **23**:861–867.

Cole ST, et al. 2001. Massive gene decay in the leprosy bacillus. *Nature* **409**:1007–1011.

Cole ST, et al. 1998. Deciphering the biology of *Mycobacterium tuberculosis* from the complete genome sequence. *Nature* **393**:537–544.

Cossart P. 2011. Illuminating the landscape of host-pathogen interactions with the bacterium *Listeria monocytogenes*. *Proc Natl Acad Sci USA* **108**:19484–19491.

Sperandio B, Fischer N, Sansonetti PJ. 2015. Mucosal physical and chemical innate barriers: lessons from microbial evasion strategies. *Semin Immunol* **27**:111–118.

Chapter 13
The Multiple Strategies of Pathogenic Bacteria

Isberg RR, Falkow S. 1985. A single genetic locus encoded by *Yersinia pseudo-tuberculosis* permits invasion of cultured animal cells by *Escherichia coli* K12. *Nature* **317**:262–264.

Galan JE, Curtiss R III. 1989. Cloning and molecular characterization of genes whose products allow *Salmonella typhimurium* to penetrate tissue culture cells. *Proc Natl Acad Sci USA* **86**:6383–6387.

Cossart P, Boquet P, Normark S, Rappuoli R. 1996. Cellular microbiology emerging. *Science* **271**:315–316.

Finlay BB, Cossart P. 1997. Exploitation of host cell functions by bacterial pathogens. *Science* **276**:718–725.

Cossart P, Sansonetti PJS. 2004. Bacterial invasion: the paradigms of enteroinvasive pathogens. *Science* **304**:242–248.

Galan JE, Cossart P. 2004. Host-pathogen interactions: a diversity of themes, a variety of molecular machines. *Curr Opin Microbiol* **8**:1–3.

Cossart P, Roy CR. 2010. Manipulation of host membrane machinery by bacterial pathogens. *Curr Opin Cell Biol* **22**:547–554.

Hubber A, Roy CR. 2010. Modulation of host cell function by *Legionella pneumophila* type IV effectors. *Annu Rev Cell Dev Biol* **26**:261–283.

Pizarro-Cerdá J, Kühbacher A, Cossart P. 2012. Entry of *Listeria* in mammalian cells: an updated view. *Cold Spring Harb Perspect Med* **2**(11):pii: a010009. doi:10.1101/cshperspect.a010009

Bierne H, Hamon M, Cossart P. 2012. Epigenetics and bacterial infections. *Cold Spring Harb Perspect Med* **2**(12):a010272. doi:10.1101/cshperspect.a010272

Puhar A, Sansonetti PJ. 2014. Type III secretion system. *Curr Biol* **24**:R84–91.

Helaine S, Cheverton AM, Watson KG, Faure LM, Matthews SA, Holden DW. 2014. Internalization of *Salmonella* by macrophages induces formation of nonreplicating persisters. *Science* **343**:204–208.

Rolando M, Buchrieser C. 2014. *Legionella pneumophila* type IV effectors hijack the transcription and translation machinery of the host cell. *Trends Cell Biol* **24**:771–778.

Arena ET, Campbell-Valois FX, Tinevez JY, Nigro G, Sachse M, Moya-Nilges M, Nothelfer K, Marteyn B, Shorte SL, Sansonetti PJ. 2015. Bioimage analysis of *Shigella* infection reveals targeting of colonic crypts. *Proc Natl Acad Sci USA* **112:**3282–3290.

Spanò S, Gao X, Hannemann S, Lara-Tejero M, Galán JE. 2016. A bacterial pathogen targets a host Rab-family GTPase defense pathway with a GAP. *Cell Host Microbe* **19:**216–226.

Chapter 14
Pathogenic Bacteria in Insects

Vallet-Gely I, Lemaitre B, Boccard F. 2008. Bacterial strategies to overcome insect defences. *Nature Rev Microbiol* **6:**302–313.

Nielsen-Leroux C, Gaudriault S, Ramarao N, Lereclus D, Givaudan A. 2012. How the insect pathogen bacteria *Bacillus thuringiensis* and *Xenorhabdus/Photorhabdus* occupy their hosts. *Curr Opin Microbiol* **15:**220–231.

Chapter 15
Plants and Their Pathogenic Bacteria

Mole BM, Baltrus DA, Dangl JL, Grant SR. 2007. Global virulence regulation networks in phytopathogenic bacteria. *Trends Microbiol* **15:**363–371.

Hogenhout SA, Oshima K, Ammar E, Kakizawa S, Kingdom H, Namba S. 2008. Phytoplasmas: bacteria that manipulate plants and insects. *Mol Plant Pathol* **9:**403–423.

Kay S, Bonas U. 2009. How *Xanthomonas* type III effectors manipulate the host plant. *Curr Opin Microbiol* **12:**37–43.

Sugio A, MacLean A, Kingdom H, Grieve VM, Manimekalia R, Hogenhout S. 2011. Diverse targets of *Phytoplasma* effectors: from plant development to defense against insects. *Annu Rev Phytopathol* **49:**175–195.

Dou D, Zhou JM. 2012. Phytopathogen effectors subverting host immunity: different foes, similar battleground. *Cell Host Microbe* **12:**484–495.

Deslandes L, Rivas S. 2012. Catch me if you can: bacterial effectors and plant targets. *Trends Plant Sci* **17:**644–655.

Chapter 16
New Visions in Infection Defense

Genetic Theory of Infectious Diseases

Casanova J-L, Abel L. 2002. Genetic dissection of immunity to bacteria: the human model. *Annu Rev Immunol* **20:**581–620.

Lam-Yuk-Tseung S, Gros P. 2003. Genetic control of susceptibility to bacterial infections in mouse models. *Cell Microbiol* **5:**299–313.

Quintana-Murci L, Alcais A, Abel L, Casanova J-L. 2007. Immunology in natura: clinical, epidemiological and evolutionary genetics of infectious diseases. *Nat Immunol* **8:**1165–1171.

Casanova J-L, Abel L. 2013. The genetic theory of infectious diseases: a brief history and selected illustrations. *Annu Rev Genomics Hum Genet* **14:**215–243.

Health Security in the Age of Globalizing Risks

World Health Organization: http://www.who.int/en/.

PART IV:
Bacteria as Tools

Chapter 17
Bacteria as Tools for Research

Restriction Enzymes

Dussoix D, Arber W. 1962. Host specificity of DNA produced by *Escherichia coli. J Mol Biol* **5:**37–49.

PCR

Saiki R, Gelfand D, Stoffel S, Scharf S, Higuchi R, Horn G, Mullis K, Erlich H. 1988. Primer-directed enzymatic amplification of DNA with a thermostable DNA polymerase. *Science* **239:**487–491.

Bacteria and Optogenetics

Oesterhelt D, Stoekenius W. 1971. Rhodopsin-like protein from the purple membrane of *Halobacterium halobium. Nat New Biol* **233:**149–152.

Williams S, Deisseroth K. 2013. Optogenetics. *Proc Natl Acad Sci USA* **110:**16287.

Deisseroth K. 2011. Optogenetics. *Nat Methods* **8:**26–29.

The CRISPR/Cas9 Revolution

Lafountaine JS, Fathe K, Smyth HDC. 2015. Delivery and therapeutic applications of gene editing technologies ZFNs, TALENs and CRISPR/Cas9. *Int J Pharmaceut* **494:**180–194.

Using Pathogenic Bacteria To Understand Eukaryotic Cells

Kocks C, Gouin E, Tabouret M, Berche P, Ohayon H, Cossart P. 1992. *Listeria monocytogenes*-induced actin assembly requires the *actA* gene, a surface protein. *Cell* **68:**521–531.

Ridley AJ, Hall A. 1992. The small GTP-binding protein rho regulates the assembly of focal adhesion and actin stress fibers in response to growth factors. *Cell* 70:389–399.

Bierne H, Cossart P. 2012. When bacteria target the nucleus: the emerging family of nucleomodulins. *Cell Microbiol* 14:622–633.

Chapter 18
Bacteria: Old and New Health Tools
Bacteria in Food

Morelli L. 2014. Yogurt, living cultures and gut health. *Am J Clin Nutr* 99: 1248S–1250S.

Probiotics

Mackowiak PA. 2013. Recycling Metchnikoff: probiotics, the intestinal microbiome and the quest for long life. *Front Publ Health* 1:1–3.

Sassone-Corsi M, Raffatelu M. 2015. No vacancy: how beneficial microbes cooperate with immunity to provide colonization resistance to pathogens. *J Immunol* 194:4081–4087.

Nami Y, Haghshenas B, Abdullah N, Barzagari A, Radiah D, Rosli R, Khostoushahi AY. 2015. Probiotics or antibiotics: future challenges in medicine. *J Med Microbiol* 64:137–146.

Fecal Transplants

Borody TJ, Khoruts A. 2011. Fecal microbiota transplantation and emerging applications. *Nature Rev Gastroenterol Hepatol* 9:88–96.

Smits LP, Bouter KE, De Vos WM, Borody TJ, Niewdorp M. 2013. Therapeutic potential of fecal microbiota transplantation. *Gastroenterology* 145:946–953.

The Intestinal Microbiota of Insect Vectors

Engel P, Moran NA. 2013. The gut microbiota of insects—diversity in structure and function. *FEMS Microbiol Rev* 37:699–735.

Hedge S, Rasgon JL, Hughes GL. 2015. The microbiome modulates arbovirus transmission in mosquitoes. *Curr Opin Virol* 15:97–102.

CRISPR/Cas9 and Gene Therapy

Sander J, Joung JK. 2014. CRISPR-Cas systems for editing, regulating and targeting genomes. *Nat Biotech* 32:347–355.

Vogel G. 2015. Bioethics. Embryo engineering alarm. *Science* 347:1301.

Baltimore D, et al. 2015. A prudent path forward for genomic engineering and germ-line modification: a framework for open discourse on the use of CRISPR-Cas9 technology to manipulate the human genome is urgently needed. *Science* **348:**36–37.

Rath D, Amlinger L, Rath A, Lundgren M. 2015. The CRISPR-Cas immune system: biology, mechanisms and applications. *Biochimie* **117:**119–128.

Bosley K, et al. 2015. CRISPR germ line engineering—the community speaks. *Nat Biotech* **33:**478–486.

Synthetic Biology

Malyshev D, Dhami K, Lavergne T, Chen T, Dai N, Foster JM, Correa I Jr, Romesberg FE. 2014. A semi synthetic organism with an expanded genetic alphabet. *Nature* **509:**385–388.

Breitling R, Takano E. 2015. Synthetic biology advances for pharmaceutical production. *Curr Opin Biotechnol* **35:**46–51.

Liu W, Stewart CN. 2015. Plant synthetic biology. *Trends Plant Sci* **20:**309–317.

Hutchison CA III, et al. 2016. Design and synthesis of a minimal bacterial genome. *Science* **351:**aad6253. doi:10.1126/science.aad6253.

Chapter 19
Bacteria as Environmental Tools

Bacillus thuringiensis as a Biopesticide

Van Frankenhuysen K. 2009. Insecticidal activity of *Bacillus thuringiensis* crystal proteins. *J Invertebr Pathol* **101:**1–16.

Pardo-López L, Soberón M, Bravo A. 2013. *Bacillus thuringiensis* insecticidal three-domain Cry toxins: mode of action, insect resistance and consequences for crop protection. *FEMS Microbiol Rev* **37:**3–22.

Bravo A, Gómez I, Porta H, García-Gómez BI, Rodriguez-Almazan C, Pardo L, Soberón M. 2013. Evolution of *Bacillus thuringiensis* Cry toxins insecticidal activity. *Microb Biotechnol* **6:**17–26.

Elleuch J, Tounsi S, Belguith Ben Hassen N, Lacrois MN, Chandre F, Jaoua S, Zghal RZ. 2015. Characterization of novel *Bacillus thuringiensis* isolates against *Aedes aegypti* (Diptera: Culicidae) and *Ceratitis capitata* (Diptera: tephridae). *J Invertebr Pathol* **124:**90–95.

Bacillus subtilis To Protect Plant Roots

Cawoy H, Mariuto M, Henry G, Fisher C, Vasileva C, Thonart N, Dommes J, Ongena M. 2014. Plant defence stimulation by natural isolates of *Bacilllus* depends on efficient surfactin production. *Mol Plant Microbe Interact* **27:**87–100.

Wolbachia and Biocontrol of Mosquito-Borne Infectious Diseases

Teixiera L, Ferreira A, Ashburner M. 2008. The bacterial symbiont *Wolbachia* induces resistance to RNA viral infections in *Drosophila melanogaster*. *PLoS Biol* 6:2753–2763.

Iturbe-Ormaetxe I, Walker T, Neill SLO. 2011. *Wolbachia* and the biological control of mosquito-borne disease. *EMBO Rep* 12:508–518.

Hoffmann AA, Montgomery BL, Popovici J, Iturbe-Ormaetxe I, Johnson PH, Muzzi F, Greenfield M, Durkan M, Leong YS, Dong YX. 2011. Successful establishment of *Wolbachia* in *Aedes* populations to suppress dengue transmission. *Nature* 476:454–457.

Fenton A, Johnson KN, Brownlie JC, Hurst GDD. 2011. Solving *Wolbachia* paradox: modeling the tripartite interaction between host, *Wolbachia* and a natural enemy. *Am Nat* 178:333–342.

Vavre F, Charlat S. 2012. Making (good) use of *Wolbachia*: what the model says. *Curr Opin Microbiol* 15:263–268.

Caragata EP, Dutra HLC, Moreira LA. 2016. Exploiting intimate relationships: controlling mosquito-transmitted disease with *Wolbachia*. *Trends Parasitol* 32:207–218.

PHOTO CREDITS

Figure 1. Eric Gaba—NASA Astrobiology Institute (public domain).

Figure 2. Institut Pasteur and Juan J. Quereda (Institut Pasteur).

Figure 3. Urs Jenal (Biozentrum, Bâle, Switzerland).

Figure 4. Institut Pasteur.

Figure 5. Juan J. Quereda (Institut Pasteur).

Figure 6. (Left) Courtesy Agnes Ullmann; (right) F. Jacob and J. Monod, *J MolBiol* **3**:318–356, 1961.

Figure 7. Jeff Mellin and Juan J. Quereda (Institut Pasteur).

Figure 8. Nina Sesto (Institut Pasteur).

Figure 9. Institut Pasteur/Antoinette Ryter.

Figure 10. http://www.servier.com/Powerpoint-image-bank and Juan J. Quereda (Institut Pasteur).

Figure 11. Juan J. Quereda (Institut Pasteur).

Figure 12. Institut Pasteur/Ashwini Chauhan, Christophe Beloin, Jean-Marc Ghigo, unité Génétique des biofilms; Brigitte Arbeille and Claude Lebos (LBCME, Faculté de Médecine, Tours, France).

Figure 13. http://www.servier.com/Powerpoint-image-bank and Juan J. Quereda (Institut Pasteur).

Figure 14. http://www.servier.com/Powerpoint-image-bank and Juan J. Quereda (Institut Pasteur).

Figure 15. http://www.servier.com/Powerpoint-image-bank and Juan J. Quereda (Institut Pasteur).

Figure 16. Institut Pasteur (unité Interactions bactéries-cellules).

Figure 17. Institut Pasteur/Marie-Christine Prévost and Agathe Subtil.

Figure 18. Édith Gouin (Institut Pasteur).

INDEX